Judy & Alan McNeilly
1991

GENE REGULATION: A EUKARYOTIC PERSPECTIVE

TITLES OF RELATED INTEREST

The eukaryote genome in development and evolution
B. John & G. Miklos

Theoretical population genetics
J. Gale

Genethics
D. Suzuki & P. Knudtson

The handling of chromosomes (6th edition)
C.D. Darlington & L.F. LaCour

Chromosomes today, volume 9
A. Stahl, J.M. Luciani & A.M. Vagner–Capodano (editors)

Chromosomes today, volume 10
K. Fredga & M.D. Bennett (editors)

Yeast biotechnology
D.R. Berry, I. Russell & G.C. Stewart (editors)

Cell movement and cell behaviour
J.M. Lackie

The physical chemistry of membranes
B. Silver

GENE REGULATION: A EUKARYOTIC PERSPECTIVE

David S. Latchman

Director, Medical Molecular Biology Unit,
University College and Middlesex School of Medicine,
London

London
UNWIN HYMAN
Boston   Sydney   Wellington

Published by the Academic Division of
Unwin Hyman Ltd
15/17 Broadwick Street, London W1V 1FP, UK

Unwin Hyman Inc.,
8 Winchester Place, Winchester, Mass. 01890, USA

Allen & Unwin (Australia) Ltd,
8 Napier Street, North Sydney, NSW 2060, Australia

Allen & Unwin (New Zealand) Ltd in association with the
Port Nicholson Press Ltd,
Compusales Building, 75 Ghuznee Street, Wellington 1,
New Zealand

First published in 1990

British Library Cataloguing in Publication Data
Latchman, David S.
Gene regulation : a eukaryotic perspective.
1. Organisms. Cells. Genes. Regulation
I. Title
574.87322

ISBN 0–04–445242–X
ISBN 0–04–445243–8

Library of Congress Cataloging-in-Publication Data
Latchman, David S.
Gene regulation : a eukaryotic perspective / David S. Latchman.
p.   cm
Includes bibliographical references.
ISBN 0–04–445242–X (alk. paper). — ISBN 0–04–445243–8 (pbk. : alk.
paper)
1. Genetic transcription. 2. Genetic regulation. 3. Eukaryotic
cells. I. Title
[DNLM: 1. Cells. 2. Gene Expression Regulation.
3. Transcription, Genetic.   QH 450.2 L351g]
QH450.2.L37   1990
574.87'3223—dc20
DNLM/DLC
for Library of Congress   90–11946
                                                    CIP

Typeset in 10 on 12 point by Columns of Reading
and printed in Great Britain by Cambridge University Press

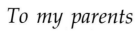

To my parents

Preface

An understanding of how genes are regulated in humans and higher eukaryotes is essential for the understanding of normal development and disease. For these reasons this process is discussed extensively in many current texts. Unfortunately, however, many of these works devote most of their space to a consideration of gene regulation in bacteria and then discuss the complexities of the eukaryotic situation more briefly. Although such an approach was the only viable one when only the simpler systems were understood, it is clear that sufficient information is now available to discuss eukaryotic gene regulation as a subject in its own right.

This work aims to provide a text which focuses on how cellular genes are regulated in eukaryotes. For this reason bacterial examples are included only to illustrate basic control processes, while examples from the eukaryotic viruses are used only when no comparable cellular gene example of a type of regulation is available.

It is hoped that this approach will appeal to a wide variety of students and others interested in cellular gene regulation. Thus, the first four chapters provide an introduction to the nature of transcriptional regulation as well as a discussion of the other levels at which genes can be regulated, which is suitable for second-year undergraduates in biology or medicine. A more extensive discussion of the details of transcriptional control, at a level suitable for final-year undergraduates, is contained in Chapters 5–7. The entire work should also provide an introduction to the topic for starting researchers, or those moving into this area from other fields and wishing to know how the gene which they are studying might be regulated. Clinicians moving into molecular research should find the concentration on the mechanisms of human cellular gene regulation of interest, and for these readers, in particular, Chapter 8 focuses on the malregulation of gene expression which occurs in human diseases and, specifically, in cancer.

I would like to thank Professor Martin Raff and Dr Paul Brickell for their critical reading of the text, and Dr Joan Heaysman who suggested that I should write this book. I am also extremely grateful to Mrs Rose Lang for typing the text and tolerating my continual changes, and to Mrs Jane Templeman who prepared the illustrations.

David S. Latchman

Acknowledgements

I would like to thank all those colleagues who have given permission for material from their papers to be reproduced in this book and have provided prints suitable for reproduction. Other than those listed below, all figures have been drawn especially for this work.

Figures 1.4 and 1.5, photographs kindly provided by Dr A. Moore, from Moore *et al.*, *Eur. J. Biochem.* **152**, 729 (1985), by permission of Springer-Verlag; Figure 1.7, redrawn from Bishop *et al.*, *Nature* **250**, 199 (1974), by permission of Dr J. Bishop and Macmillan Magazines Ltd; Figure 1.8, redrawn from Hastie & Bishop, *Cell* **9**, 761 (1976), by permission of Dr J. Bishop and Cell Press; Figure 1.9, from Latchman *et al.*, *Biochim. Biophys. Acta* **783**, 130 (1984), by permission of Elsevier Science Publishers; Figure 1.10, photograph kindly provided by Dr P. Brickell, from Devlin *et al.*, *Development* **103**, 111 (1988), by permission of the Company of Biologists Ltd.

Figure 2.2, photograph kindly provided by Dr A.T. Sumner; Figure 2.4, redrawn from Cove, D.J., *Genetics* (1971), by permission of Professor D.J. Cove and Cambridge University Press; Figure 2.5, redrawn from Steward *et al.* *Science* **143**, 20 (1964), by permission of Professor F.C. Steward and the American Association for the Advancement of Science (AAAS); Figure 2.6, redrawn from Gurdon, *Gene Expression* (1974), by permission of Professor J.B. Gurdon and Oxford University Press; Figure 2.10, photograph kindly provided by Dr M. Ashburner; Figure 2.13, photograph kindly provided by Dr R. Manning, from Manning & Gage, *J. Biol. Chem.* **253**, 2044 (1978); Figure 2.14, redrawn from Spradling & Mahowald, *PNAS* **77**, 1096 (1980), by permission of Dr A. Spradling; Figure 2.19, redrawn from Hicks, *Nature* **326**, 445 (1987), by permission of Dr J. Hicks and Macmillan Magazines Ltd.

Figures 3.3. and 3.4, photographs kindly provided by Professor B.W. O'Malley, from Roop *et al.*, *Cell* **15**, 671 (1978), by permission of Cell Press; Figure 3.9, photograph kindly provided by Dr M. Ashburner; Figure 3.11, photograph kindly provided by Dr R.S. Hill, from Hill & Macgregor, *J. Cell Sci.* **44**, 87 (1980), by permission of the Company of Biologists Ltd.

Figure 4.10, redrawn from Breitbart & Nadal-Ginard, *Cell* **49**, 793 (1987), by permission of Professor B. Nadal-Ginard and Cell Press; Figure 4.12, redrawn from Guyette *et al.*, *Cell* **17**, 1013 (1979), by permission of Professor J. Rosen and Cell Press, Figure 4.13, redrawn from Casey *et al.*, *Science* **240**, 924 (1988), by permission of Dr J.B. Harford and AAAS; Figure 4.17, photograph kindly provided by Dr N. Standart and Dr T. Hunt.

Figure 5.3, redrawn from Gurdon, J.B., *Gene Expression* (1974), by permission of Professor J.B. Gurdon and Oxford University Press; Figure 5.4, redrawn from Cove D.J., *Genetics* (1971), by permission of Professor D.J. Cove and

Cambridge University Press; Figure 5.6, photographs kindly provided by Dr J.T. Finch, from Finch *et al.*, *PNAS* **72**, 3320 (1975); Figure 5.7, photograph kindly provided by Dr F. Thoma, from Thoma *et al.*, *J. Cell. Biol.* **83**, 403 (1979), by permission of Rockefeller Press; Figure 5.8, photograph kindly provided by Dr J. McGhee, from McGhee *et al.*, *Cell* **33**, 831 (1983), by permission of Cell Press; Figure 5.9, photograph kindly provided by Professor O.L. Miller, from McKnight *et al.*, *Cold Spring Harbor Symposium* **42**, 741 (1978), by permission of Cold Spring Harbor Laboratory; Figure 5.14, redrawn from Weintraub *et al.*, *Cell* **24**, 333 (1981), by permission of Professor H. Weintraub and Cell Press; Figure 5.17, redrawn from Keshet *et al.*, *Cell* **44**, 535 (1986), by permission of Professor H. Cedar and Cell Press; Figure 5.20, photograph kindly provided by Professor P. Chambon, from Kaye *et al.*, *EMBO Journal* **3**, 1137 (1984), by permission of Oxford University Press; Figure 5.22, redrawn from Nordheim & Rich, *Nature* **303**, 674 (1983), by permission of Professor A. Rich and Macmillan Magazines Ltd; Figure 5.23, photograph kindly provided by Professor M. Yaniv, from Saragosti *et al.*, *Cell* **20**, 65 (1980), by permission of Cell Press.

Figures 6.10 and 6.11, photographs kindly provided by Professor D. Hanahan, from Hanahan, *Nature* **315**, 115 (1985), by permission of Macmillan Magazines Ltd.; Figure 6.19, redrawn from Murphy *et al.*, *Cell* **35**, 865 (1983), by permission of Dr P.W.J. Rigby and Cell Press; Figure 6.20, photograph kindly provided by Professor W.E. Hahn, from Owens *et al.*, *Science* **229**, 1263 (1985), by permission of AAAS; Figures 6.24 and 6.26, photographs kindly provided by Professor D.D. Brown, 6.24 from Sakonju *et al.*, *Cell* **19**, 13 (1980), 6.26 from Sakonju & Brown, *Cell* **31**, 395 (1982), by permission of Cell Press.

Figure 7.4, photograph kindly provided by Professor W.J. Gehring, from Gehring, *Science* **236**, 1245 (1987), by permission of AAAS; Figure 7.10, redrawn from Schleif, *Science* **241**, 1182 (1988), by permission of Dr R. Schleif and AAAS; Figure 7.12, redrawn from Evans & Hollenberg, *Cell* **52**, 1 (1988), by permission of Professor R.M. Evans and Cell Press; Figure 7.13 redrawn from Klug & Rhodes, *Trends in Biochemical Science* **12**, 464 (1987), by permission of Professor Sir Aaron Klug and Elsevier Publications; Figure 7.21, redrawn from Turner & Tjian, *Science* **243**, 1189 (1989), by permission of Professor R. Tjian and AAAS; Figure 7.26, redrawn from Giniger & Ptashne, *Nature* **330**, 670 (1987), by permission of Professor M. Ptashne and Macmillan Magazines Ltd.

Figure 8.3, redrawn from Takeya & Hanafusa, *Cell* **32**, 881 (1985), by permission of Professor H. Hanafusa and Cell Press.

Contents

Preface   *Page*   ix
Acknowledgements   xi
List of tables   xvii

1 *Tissue-specific expression of proteins and messenger RNAs*   1
   1.1   Introduction   1
   1.2   Tissue-specific expression of proteins   1
       1.2.1   General methods for studying the protein
           composition of tissues   2
       1.2.2   Specific methods for studying the protein
           composition of tissues   4
   1.3   Tissue-specific expression of messenger RNAs   6
       1.3.1   General methods for studying the mRNAs
           expressed in different tissues   7
       1.3.2   Specific methods for studying the mRNAs
           expressed in different tissues   9
   1.4   Conclusions   12
       References   12

2 *The DNA of different cell types is similar in both amount*
   and type   13

   2.1   Introduction   13
   2.2   DNA loss   13
       2.2.1   DNA loss as a mechanism of gene regulation   13
       2.2.2   Chromosomal studies   14
       2.2.3   Functional studies   19
       2.2.4   Molecular studies   22
   2.3   DNA amplification   24
       2.3.1   DNA amplification as a mechanism of gene
           regulation   24
       2.3.2   Chromosomal studies   25
       2.3.3   Molecular studies   28
   2.4   DNA rearrangement   30
   2.5   Conclusions   43
       References   44

CONTENTS

3 *Regulation at transcription*   46

    3.1 Levels of gene regulation   46
    3.2 Evidence for transcriptional regulation   48
         3.2.1 Evidence from studies of nuclear RNA   48
         3.2.2 Evidence from pulse-labelling studies   53
         3.2.3 Evidence from nuclear run-on assays   56
         3.2.4 Evidence from polytene chromosomes   59
    3.3 Regulation at transcriptional elongation   61
    3.4 Conclusions   65
         References   65

4 *Post-transcriptional regulation*   67

    4.1 Regulation after transcription?   67
    4.2 Regulation of RNA splicing   68
         4.2.1 RNA splicing   68
         4.2.2 Alternative RNA splicing   69
         4.2.3 Mechanism of alternative RNA splicing   78
         4.2.4 Generality of alternative RNA splicing   81
    4.3 Regulation of RNA transport   82
    4.4 Regulation of RNA stability   83
         4.4.1 Cases of regulation by alterations in RNA stability   83
         4.4.2 Mechanisms of stability regulation   85
         4.4.3 Role of stability changes in regulation of
                gene expression   89
    4.5 Regulation of translation   90
         4.5.1 Cases of translational control   90
         4.5.2 Mechanism of translational control   91
         4.5.3 Significance of translational control   96
    4.6 Conclusions   97
         References   97

5 *Transcriptional control – chromatin structure*   100

    5.1 Introduction   100
    5.2 Commitment to the differentiated state and its stability   103
    5.3 Chromatin structure   106
    5.4 Changes in chromatin structure in active or potentially
         active genes   112
         5.4.1 Active DNA is organized in a nucleosomal structure   112
         5.4.2 Sensitivity of active chromatin to DNaseI digestion   113
         5.4.3 Mechanism of increased DNaseI sensitivity   116
    5.5 Other changes in DNA and its associated proteins in
         active or potentially active genes   118
         5.5.1 DNA methylation   118
         5.5.2 Histone modifications   125

5.6 DNaseI hypersensitive sites in active or potentially
    active genes   126
    5.6.1 Detection of DNaseI hypersensitive sites   126
    5.6.2 Nature and significance of hypersensitive sites   131
5.7 Conclusions   134
    References   135

6 *Transcriptional control – DNA sequence elements*   138

6.1 Introduction   138
    6.1.1 Relationship of gene regulation in
        prokaryotes and eukaryotes   138
    6.1.2 Complexity of the eukaryotic system   138
    6.1.3 The Britten and Davidson model for the co-ordinate
        regulation of unlinked genes   141
6.2 Short sequence elements located within or adjacent
    to the gene promoter   142
    6.2.1 Short regulatory elements   142
    6.2.2 The heat-shock response element   143
    6.2.3 Other response elements   148
    6.2.4 Mechanism of action of promoter regulatory
        elements   151
6.3 Enhancers   153
    6.3.1 Regulatory sequences that act at a distance   153
    6.3.2 Tissue-specific activity of enhancers   155
    6.3.3 Mechanism of action of enhancers   158
    6.3.4 Positive and negative action of enhancer
        elements   161
6.4 Role of repeated sequences   164
    6.4.1 Repeated DNA   164
    6.4.2 Role of repeated sequences in gene expression   165
6.5 Regulation of transcription by RNA polymerases I
    and III   170
    6.5.1 RNA polymerase I   170
    6.5.2 RNA polymerase III   170
6.6 Conclusions   176
    References   176

7 *Transcriptional control – transcription factors*   180

7.1 Introduction   180
7.2 DNA binding by transcription factors   185
    7.2.1 Introduction   185
    7.2.2 The helix-turn-helix motif   185
    7.2.3 The zinc finger motif   194

7.2.4	The leucine zipper and the basic DNA-binding domain	204
7.2.5	Other DNA-binding domains	207
7.3	Activation of transcription	207
7.3.1	Introduction	207
7.3.2	Activation domains	208
7.3.3	How is transcription activated?	213
7.4	What activates the activators?	217
7.4.1	Introduction	217
7.4.2	Regulated synthesis of transcription factors	219
7.4.3	Regulated activity of transcription factors	221
7.5	Conclusions	225
	References	227

8 Gene regulation and cancer — 231

8.1	Introduction	231
8.2	Proto-oncogenes	231
8.3	Elevated expression of oncogenes	237
8.4	Transcription factors as oncogenes	243
8.4.1	Fos, Jun, and AP1	243
8.4.2	v-erbA and the thyroid hormone receptor	246
8.4.3	Other transcription-factor-related oncogenes	251
8.5	Conclusions	252
	References	254

9 Conclusions and future prospects — 257

	References	259

Index — 260

List of tables

4.1 Cases of alternative splicing which are regulated developmentally or tissue specifically   Page 72/3

4.2 Regulation of RNA stability   84

4.3 Regulation of the transferrin receptor and ferritin genes   94

5.1 The histones   107

5.2 Examples of genes containing DNaseI hypersensitive sites   129

6.1 Eukaryotic RNA polymerases   139

6.2 Sequences present in the upstream region of the *hsp70* gene which are also found in other genes   144

6.3 Sequences that confer response to a particular stimulus   149

6.4 Relationship of consensus sequences conferring responsivity to various hormones   150

7.1 Transcriptional regulatory proteins containing Cys_2His_2 zinc fingers   197

7.2 Transcriptional regulatory proteins with multiple cysteine fingers   200

7.3 Transcription factor domains   226

8.1 Proto-oncogenes and their functions   234

CHAPTER ONE

Tissue-specific expression of proteins and messenger RNAs

1.1 INTRODUCTION

The evidence that eukaryotic gene expression must be a highly regulated process is available to anyone visiting a butcher's shop. The various parts of the mammalian body on display differ dramatically in appearance, ranging from the muscular legs and hind quarters to the soft tissues of the kidneys and liver. However, all these diverse types of tissues arose from a single cell, the fertilized egg or zygote, raising the question of how this diversity is achieved. It is the aim of this book to consider the processes regulating tissue-specific gene expression in mammals and the manner in which they produce these differences in the nature and function of different tissues.

1.2 TISSUE-SPECIFIC EXPRESSION OF PROTEINS

The fundamental dogma of molecular biology is that DNA produces RNA which in turn produces proteins. Thus the genetic information in the DNA specifying particular functions is converted into an RNA copy, which is then translated into protein. The action of the protein then produces the phenotype, be it the presence of a functional globin protein transporting oxygen in the blood or the activity of a proteinaceous enzyme capable of producing the pigment causing the appearance of brown rather than blue eyes. Hence, if the differences in the appearance of mammalian tissues described above are indeed caused by differences in gene expression in different tissues, they should be produced by differences in the proteins present in these tissues. Such differences can be detected both by general methods

1

aimed at studying the expression of all proteins in a given tissue and by specific methods which study the expression of one particular protein.

1.2.1 *General methods for studying the protein composition of tissues*

If the proteins in a cell are denatured by treatment with the detergent sodium dodecyl sulphate (SDS) and then subjected to electrophoresis in a polyacrylamide gel (Laemmli 1970), they can be visualized by staining the gel. This method allows the demonstration of some differences between different cells and tissues (Fig. 1.1). Because of the very large number of proteins in the cell and the limited resolution of the technique, one-dimensional gel electrophoresis cannot be used to extensively investigate the variation of proteins between tissues. Thus, for example, two entirely different proteins in two tissues may be scored as being the same protein simply on the basis of a similarity in

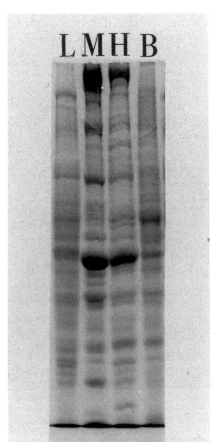

Figure 1.1 One-dimensional SDS-polyacrylamide gel electrophoresis of cellular proteins isolated from mouse brain (track B), mouse heart (track H), mouse liver (track L), and mouse skeletal muscle (track M).

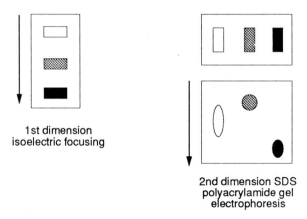

1st dimension
isoelectric focusing

2nd dimension SDS
polyacrylamide gel
electrophoresis

Figure 1.2   Two-dimensional gel electrophoresis.

size. A more detailed investigation of the protein composition of different tissues can be achieved by two-dimensional gel electrophoresis (O'Farrell 1975). In this procedure (Fig. 1.2), proteins are first separated on the basis of differences in their charge, in a technique known as isoelectric focusing, and the separated proteins, still in the first gel, are layered on top of an SDS-polyacrylamide gel. In the subsequent electrophoresis the proteins are separated by their size. Hence a protein moves to a position determined both by its size and its charge. The much greater resolution of this method allows a number of differences in the protein composition of particular tissues to be identified. Thus some spots or proteins are found in only one or a few tissues and not in many others, while others are found at dramatically different abundance in different tissues (Fig. 1.3). Hence, the different

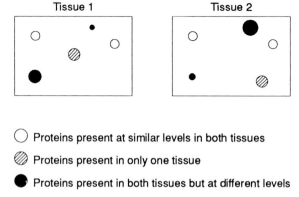

Tissue 1   Tissue 2

◯ Proteins present at similar levels in both tissues

▨ Proteins present in only one tissue

● Proteins present in both tissues but at different levels

Figure 1.3 Schematic results of two-dimensional gel electrophoresis allowing the detection of proteins specific to one tissue or expressed at different levels in different tissues.

appearances of different tissues are indeed paralleled by both qualitative and quantitative differences in the proteins present in each tissue. It should be noted, however, that some proteins can be shown by two-dimensional gel electrophoresis to be present at similar levels in virtually all tissues. Presumably such so-called housekeeping proteins are involved in basic metabolic processes common to all cell types.

1.2.2 Specific methods for studying the protein composition of tissues

The general conclusion that qualitative and quantitative differences exist in the protein composition of different tissues can also be reached by investigating the expression of particular individual proteins in such tissues. These methods might involve the isolation of widely differing amounts of an individual protein (or none of the protein at all) from different tissues, using an established purification procedure, or the detection of an enzymatic activity associated with the protein in extracts of only one particular tissue. We shall consider in detail, however, only methods where the expression of a specific protein is monitored by the use of a specific antibody to it. Normally, such an antibody is produced by injecting the protein into an animal such as a rabbit or mouse. The resulting immune response results in the presence in the animal's blood of antibodies which specifically recognize the protein and can be used to monitor its expression in particular tissues.

Such antibodies can be used in conjunction with the one-dimensional polyacrylamide gel electrophoresis technique already described (see Section 1.2.1) to investigate the expression of a particular protein in different tissues. In this technique, known as Western blotting (Gershoni & Palade 1983), the gel-separated proteins are transferred to a nitrocellulose filter which is incubated with the antibody. The antibody reacts specifically with the protein against which it is directed and which will be present at a particular position on the filter, dependent on how far it moved in the electrophoresis step and hence on its size. The binding of the antibody is then visualized by a radioactive or enzymatic detection procedure. If a tissue contains the protein of interest, a band will be observed in the track containing total protein from that tissue and the intensity of the band observed will provide a measure of the amount of protein present in the tissue. If none of the particular protein is present in a given tissue, no band will form (Fig. 1.4). Hence this method allows the presence or absence of a specific protein in a particular tissue to be assessed using one-dimensional gel electrophoresis without the complicating effect of other unrelated proteins of similar size, since these will fail to bind the antibody.

4

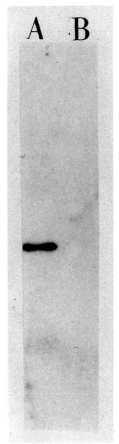

Figure 1.4 Western blot with an antibody to guinea-pig casein kinase showing the presence of the protein in lactating mammary gland (A) but not in liver (B).

The specific reaction of a protein with an antibody can also be used directly to investigate its expression within a particular tissue. In this method, known as immunofluorescence, thin sections of the tissue of interest are reacted with the antibody, which binds to those cell types expressing the protein. As before, the position of the antibody is visualized by an enzymatic detection procedure, or more usually, by the use of a fluorescent dye which can be seen in a microscope when appropriate filters are used (Fig. 1.5). Therefore this method can be used not only to provide information about the tissues expressing a particular protein but, since individual cells can be examined, it also allows detection of the specific cell types within the tissue which are expressing the protein.

As with the general methods for studying all proteins, specific methods aimed at studying the expression of particular proteins indicate that while some proteins are present at similar abundance in all tissues, others are present at widely different abundances in different tissues, and a large number are specific to one or a few tissues

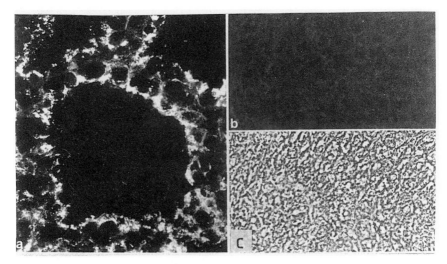

Figure 1.5 Use of the antibody to casein kinase to show the presence of the protein (bright areas) in frozen sections of lactating mammary gland (a) but not in liver (b). (c) A phase-contrast photomicrograph of the liver section, confirming that the lack of staining with the antibody in (b) is not due to the absence of liver cells in the sample.

or cell types. Hence the differences between different tissues in appearance and function are correlated with qualitative and quantitative differences in protein composition, and it is necessary to understand how such differences are produced.

1.3 TISSUE-SPECIFIC EXPRESSION OF MESSENGER RNAs

Proteins are produced by the translation of specific messenger RNA molecules on the ribosome. Hence, having established that quantitative and qualitative differences exist in the protein composition of different tissues, it is necessary to ask whether such differences are paralleled by tissue-specific differences in the abundance of their corresponding mRNAs. Thus, although it seems likely that differences in the mRNA populations of different tissues do indeed underlie the observed differences in proteins, it is possible that all tissues have the same mRNA species and that production of different proteins is controlled by regulating which of these are selected by the ribosome to be translated into protein.

As with the study of proteins, both general and specific techniques exist for studying the mRNAs expressed in a given tissue.

1.3.1 General methods for studying the mRNAs expressed in different tissues

The most commonly used method for studying RNA populations is the R_ot curve (Bishop *et al.* 1974). In this method a complementary DNA (cDNA copy) of the RNA is first made, using radioactive precursors so that the cDNA is made radioactively labelled. After dissociating the cDNA and the RNA, the rate at which the radioactive cDNA reanneals or hybridizes to the RNA is followed. In this process, more molecules

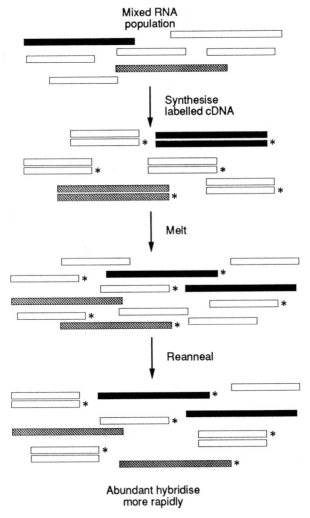

Figure 1.6 Hybridization of messenger RNA and its radioactive complementary DNA copy (indicated by the stars) in a R_ot curve experiment. The abundant RNA finds its partner and hybridizes more rapidly.

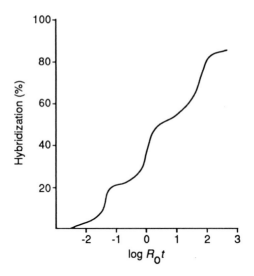

Figure 1.7 Typical R_0t curve, produced using the mRNA of a single tissue. Note the three phases of the curve, produced by hybridization of the highly abundant, moderately abundant, and rare RNA species.

of an abundant RNA will be present than of an RNA which is less abundant. The cDNA from an abundant RNA will find a partner more rapidly than one derived from a rare one, simply because more potential partners are available. Hence, when used with the RNA of one tissue, the rate of hybridization or reannealing of any cDNA molecule provides a measure of the abundance of its corresponding RNA in that tissue. The higher the abundance of a particular RNA species, the more rapidly it anneals to its complementary cDNA (Fig. 1.6). Typical R_0t curves, showing three predominant abundance classes of RNA in any given tissue, can be constructed readily (Fig. 1.7).

This method can be extended to provide a comparison of the RNA populations of two different tissues (Hastie & Bishop 1976). In this method the RNA of one tissue is mixed with labelled cDNA prepared from the RNA of another tissue and the mixture annealed. Clearly, the rate of hybridization of the radiolabelled cDNA will be determined by the abundance of its corresponding RNA in the tissue from which the RNA is derived. If the RNA is absent altogether in this tissue (being specific to that from which the cDNA is derived), no reannealing will occur. If the corresponding RNA is present, the rate at which the cDNA hybridizes will be determined by its abundance in the tissue from which the RNA is derived. Hence, by carrying out an inter-tissue R_0t curve, both qualitative and quantitative differences between the RNA populations of different tissues can be observed. Such experi-

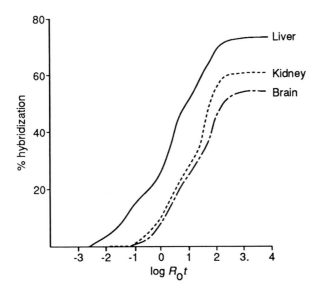

Figure 1.8 Inter-tissue *Rôt* curve, in which mRNA from the liver has been hybridized to cDNA prepared from liver, kidney, or brain mRNA. Note the lower extent of hybridization, with the kidney and brain samples indicating the absence or lower abundance of some liver RNAs in these tissues.

ments (see Figure 1.8) lead to the conclusion that, although sharing of rare RNAs (presumably encoding housekeeping proteins) does occur, both qualitative and quantitative differences between the RNA populations of different tissues are observed, with RNA species that are abundant in one cell type being rare or absent in other tissues. Thus, in a comparison of kidney and liver messenger RNAs, Hastie & Bishop (1976) found that while between 9500 and 10 500 of the 11 000 rare messenger RNA species found in kidney were also found in liver, the six highly abundant RNA species found in the kidney were either absent or present at very low levels in the liver.

1.3.2 Specific methods for studying the mRNAs expressed in different tissues

The conclusion from $R_o t$ curve analysis, that both qualitative and quantitative differences exist between RNA populations, can also be reached by the use of specific methods that detect one specific mRNA using a cloned DNA probe derived from its corresponding gene. In the most commonly used of such methods, Northern blotting (Thomas 1980), the RNA extracted from a particular tissue is electrophoresed on an agarose gel, transferred to a nitrocellulose filter, and hybridized to a radioactive probe derived from the gene encoding the mRNA of

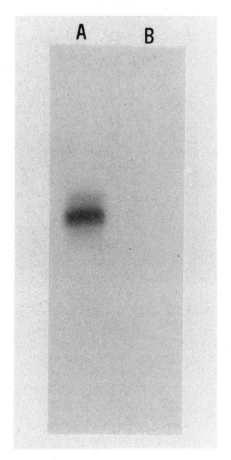

Figure 1.9 Northern blot hybridization using a probe specific for the α-foetoprotein mRNA. The RNA is detectable in the embryonic yolk sac sample (track A) but not in the adult liver sample (track B).

interest. The presence of the RNA in a particular tissue will result in binding of the radioactive probe and the visualization of a band on autoradiography, the intensity of the band being dependent on the amount of RNA present (Fig. 1.9). Although such experiments do detect the RNA encoding some proteins, such as actin or tubulin, in all tissues, very many others are found only in one particular tissue. Thus the RNA for the globin protein is found only in reticulocytes, that for myosin only in muscle, while (in the example shown in Fig. 1.9) the mRNA encoding the foetal protein, α-foetoprotein, is shown to be present only in the embryonic yolk sac and not in the adult liver.

As with protein studies, methods studying expression in RNA isolated from particular tissues can be supplemented by methods allowing direct visualization of the RNA in particular cell types. In such a method, known as *in situ* hybridization, a radioactive probe specific for the RNA to be detected is hybridized to a section of the tissue of interest in which cellular morphology has been maintained. Visualiza-

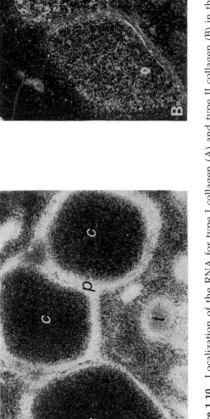

Figure 1.10 Localization of the RNA for type I collagen (A) and type II collagen (B) in the 10-day chick embryo leg by *in situ* hybridization. Note the different distributions of the bright areas produced by binding of each probe to its specific mRNA. c, cartilage; t, tendons; p, perichondrium.

tion of the position at which the radioactive probe has bound (Fig. 1.10) not only allows an assessment of whether the particular tissue is expressing the RNA of interest but also of which individual cell types within the tissue are responsible for such expression. This technique has been used to show that the expression of some mRNAs is confined to only one cell type, paralleling the expression of the corresponding proteins by that cell type.

1.4 CONCLUSIONS

It is clear, therefore, that the qualitative and quantitative variation in protein composition between different tissues is paralleled by a similar variation in the nature of the mRNA species present in different tissues. Hence, the production of different proteins by different tissues is not regulated by selecting which of a common pool of RNA molecules is selected for translation by the ribosome, although some isolated examples of such regulated translation may exist (see Ch. 4). In order to understand the basis for such tissue-specific variation in protein and mRNA composition, it is therefore necessary to move one stage further back and examine the nature of the DNA encoding these mRNAs in different tissues.

REFERENCES

Bishop, J. O., J. G. Morton, M. Rosbash, & M. Richardson, 1974. Three abundance classes in HeLa cell messenger RNA. *Nature* **250**, 199–204.
Gershoni, J. M. & G. E. Palade, 1983. Protein blotting: principles and applications. *Analytical Biochemistry* **131**, 1–15.
Hastie, N. B. & J. O. Bishop, 1976. The expression of three abundance classes of messenger RNA in mouse tissues. *Cell* **9**, 761–74.
Laemmli, U. K. 1970. Cleavage of structural proteins during the assembly of the head of the bacteriophage T4. *Nature* **227**, 680–5.
O'Farrell, P. H. 1975. High resolution two-dimensional electrophoresis of proteins. *Journal of Biological Chemistry* **250**, 4007–21.
Thomas, P. S. 1980. Hybridization of denatured RNA and small DNA fragments transferred to nitrocellulose. *Proceedings of the National Academy of Sciences of the USA* **77**, 5201–5.

Erythroblast

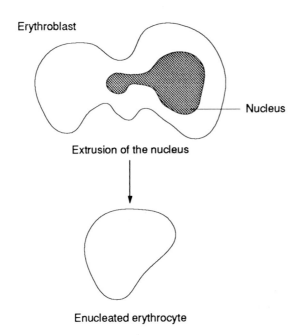

Nucleus

Extrusion of the nucleus

Enucleated erythrocyte

Figure 2.3 Extrusion of the nucleus from an erythroblast, resulting in an anucleate erythrocyte.

LOSS OF SPECIFIC CHROMOSOMES IN SOMATIC CELLS

The reticulocyte case does not, therefore, provide us with an example of selective DNA loss and, indeed, no such case has as yet been reported in mammalian cells. However, a number of such cases in which specific chromosomes are lost during development have been described in other eukaryotes, including some nematodes, crustaceans, and insects (Wilson 1928). In these cases a clear distinction is made early in embryonic development between the cells which will produce the body of the organism (the somatic cells) and those which will eventually form its gonads and allow the organism to reproduce (the germ cells). Early on in the development of the embryo, during the cleavage stage, the somatic cells lose certain chromosomes, with only the germ cells, which will give rise to subsequent generations, retaining the entire genome. Such losses can be quite extensive, for example in the gall midge *Miastor* only 12 of the total 48 chromosomes are retained in the somatic cells. It is assumed that the lost DNA is only required for some aspects of germ cell development. It is known to be rich in the highly repeated (or satellite) type of DNA, which is thought to be involved in homologous chromosome pairing during the meiotic division that occurs in germ cell development (John & Miklos 1979). Hence it is more economical for the organism to dispense with such

17

DNA in the development of somatic tissue, rather than to waste energy replicating it and passing it on to all somatic cells where it has no function.

Although the mechanism by which this selective loss of DNA occurs is not yet fully understood, there is evidence that it is mediated by cytoplasmic differences between regions of the fertilized egg or zygote. Thus, in the nematode worm *Parascaris* the distinction between somatic and germ cells is established by the first division of the zygote, the smaller of the two cells produced losing chromosomes and giving rise to the soma while the larger retains a full chromosome complement and gives rise to the germ line (Fig. 2.4). If, however, the zygote is centrifuged, the plane of the first division can be altered, resulting in the production of two similar-sized cells neither of which loses chromosomes. Hence the centrifugation process has abolished some heterogeneity in the cytoplasm responsible for this effect. Interestingly, although both daughter cells retain a full complement of chromosomes, their inability to lose chromosomes prevents normal embryonic development from occurring.

Although cases of DNA loss observable by examination of the chromosomes have been observed in some organisms, it must be emphasized that in the vast majority of cases no such loss of DNA is observed by this means. However, because the amount of DNA in a

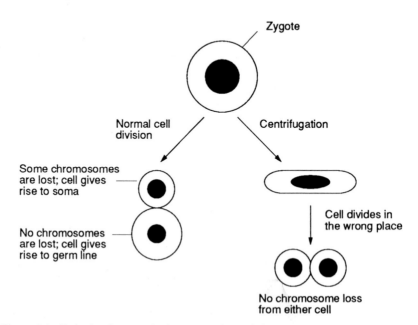

Figure 2.4 Early development in the nematode, and the consequences of changing the plane of the first cell division by centrifugation of the zygote.

Giemsa-stained band is very large (about 10 million base pairs), it is possible that losses of short regulatory regions necessary for gene expression, or losses of individual genes, could occur and not be observed by this technique. The large-scale losses seen in chromosome studies would then represent the tip of the iceberg in which varying degrees of gene loss occur in different situations. A consideration of this possibility requires the use of more sensitive methods to search for gene deletions. These involve both functional and molecular studies.

2.2.3 Functional studies

A model in which selective gene loss controls differentiation must view differentiation as essentially an irreversible process. Thus, for example, once the gene for myosin has been eliminated from an antibody-producing B-lymphocyte it will not be possible for such a cell, whatever the circumstances, to give rise to a myosin-containing muscle cell. Although the requirement for such a change from one differentiated cell type to another is not likely to occur in normal development, such a phenomenon, known as transdifferentiation, has been achieved experimentally in Amphibia (Yamada 1967). Thus, if the lens is surgically removed from an eye of one of these organisms, some of the neighbouring cells in the iris epithelium lose their differentiated phenotype and begin to proliferate. Eventually these cells differentiate into typical lens cells and produce large quantities of the lens-specific proteins, such as the crystallins, which are readily detectable with appropriate antibodies. The genes for the lens proteins must therefore be intact within the iris cells even though in normal development such genes would never be required in these cells.

The case against selective loss of DNA is strengthened still further by the very dramatic experiments in which a whole new organism has been produced from a single differentiated cell. Although not yet achieved in mammalian cells, this has been achieved in both plants (Steward 1970) and Amphibia (Gurdon 1968).

In plants such regeneration has been achieved in several species, including both the carrot and tobacco plants. In the carrot, for example, regeneration can occur from a single differentiated phloem cell, which forms part of the tubing system by which nutrients are transported in the plant. Thus if a piece of root tissue is placed in culture (Fig. 2.5), single quiescent cells of the phloem can be stimulated to grow and divide, and an undifferentiated callus-type tissue forms. The disorganized cell mass can be maintained in culture indefinitely but if the medium is suitably supplemented at various stages, embryonic development will occur and eventually result in a fully functional flowering plant, containing all the types of differentiated cells and

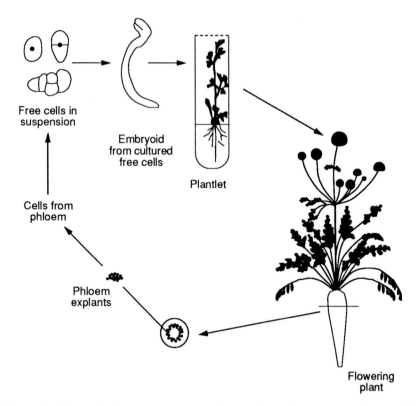

Free cells in
suspension

Embryoid
from cultured
free cells

Plantlet

Cells from
phloem

Phloem
explants

Flowering
plant

Figure 2.5 Scheme for the production of a fertile carrot plant from a single differentiated cell of the adult plant phloem, by growth in culture as a free cell suspension which develops into an embryo and then into an adult plant.

tissues normally found. The plant that forms is fertile and cannot be distinguished from a plant produced by normal biological processes.

The ability to regenerate a functional plant from fully differentiated cells eliminates the possibility that genes required in other cell types are eliminated in the course of plant development. It is noteworthy, however, that in this case, as with lens regeneration, one differentiated cell type does not transmute directly into another, rather a transitional state of undifferentiated proliferating cells serves as an intermediate. Hence although differentiation does not apparently involve permanent irreversible changes in the DNA, it appears to be relatively stable and, although reversible, requires an intermediate stage for changes to occur. This semi-stability of cellular differentiation will be discussed further in Chapter 5.

Although no complex animal has been regenerated by culturing a single differentiated cell in the manner used for plants, other techniques have been used to show that differentiated cell nuclei are capable of giving rise to very many different cell types (Gurdon 1968). These experiments involve the use of nuclear transplantation (Fig. 2.6). In this technique the nucleus of an unfertilized frog egg is destroyed, either surgically or by irradiation with ultraviolet light, and a donor nucleus from a differentiated cell of a genetically distinguishable strain of frog is implanted. Development is then allowed to proceed in order

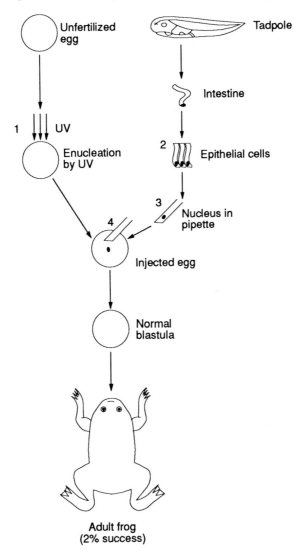

Figure 2.6 Nuclear transplantation in Amphibia. The introduction of a donor nucleus from a differentiated cell into a recipient egg whose nucleus has been destroyed by irradiation with ultraviolet light can result in an adult frog with the genetic characteristics of the donor nucleus.

to test whether, in the environment of the egg cytoplasm, the genetic information in the nucleus of the differentiated cell can produce an adult frog. When the donor nucleus is derived, for example, from differentiated intestinal epithelial cells of a tadpole, an adult frog is indeed produced in a small proportion of cases (about 2 per cent) and development to at least a normal swimming tadpole occurs in about 20 per cent of cases, with the organisms having the genetic characteristics of the donor nucleus. Such successful development supported by the nucleus of a differentiated frog cell is not unique to intestinal cells and has been achieved with the nuclei of other cell types, such as the skin cells of an adult frog. Hence although these experiments are technically difficult and have a high failure rate, it is possible for the nuclei of differentiated cells specialized for the absorption of food to produce organisms containing a range of diffferent cell types and which are perfectly normal and fertile.

As with plants, it is clear that in animals selective loss of DNA is not a general mechanism by which gene control is achieved.

2.2.4 Molecular studies

The conclusions of functional studies carried out in the 1960s and early 1970s were confirmed abundantly in the late 1970s and 1980s by molecular investigations involving the use of specific DNA probes derived from individual genes. Such probes can be used in Southern blotting experiments (Southern 1975) to investigate the structure of a particular gene in different tissues. In this technique (Fig. 2.7) DNA from a particular tissue is digested into specific fragments with restriction endonucleases which cut the DNA at particular points. Following electrophoresis on an agarose gel, the cut DNA is transferred to nitrocellulose and hybridized with a radioactive probe for the gene being studied. The structure of the gene can then be

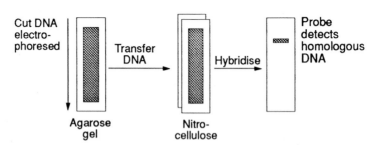

Figure 2.7 Procedure for Southern blot analysis, involving electrophoresis of DNA which has been cut with a restriction enzyme, its transfer to a nitrocellulose filter, and hybridization of the filter with a specific DNA probe for the gene of interest.

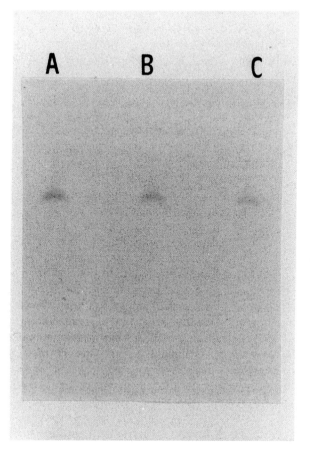

Figure 2.8 Southern blot hybridization of a radioactive probe specific for the α-foetoprotein gene to *Eco*RI-cut DNA prepared from foetal liver (track A), adult liver (track B), and adult brain (track C). Note the presence of an identically sized gene fragment in each tissue, although the gene is only expressed in the foetal liver.

examined by determining the sizes of fragments observed when the DNA is cut with particular restriction enzymes.

As with the use of Northern blotting and Western blotting (see Ch. 1) to study, respectively, the RNA and protein content of different tissues, Southern blotting can be used to investigate the structure of the DNA in individual tissues. When this is done, no losses of particular genes are observed in specific tissues. Thus, the gene for β-globin is clearly present in the DNA of brain or spleen where it is never expressed, as well as in the DNA of erythroid tissues where expression occurs (Jeffreys & Flavell 1977) and numerous other examples have been reported (Fig. 2.8).

These techniques, which could detect losses of less than 100 bases (as opposed to the millions of bases resolvable by chromosomal techniques), have now been supplemented by direct determination of the DNA sequence of particular genes in expressing and non-expressing tissues. Such techniques, which could detect the loss of even a single base of DNA, have confirmed the conclusions of all other studies, indicating that selective DNA loss is not a general mechanism of gene regulation.

2.3 DNA AMPLIFICATION

2.3.1 DNA amplification as a mechanism of gene regulation

Having eliminated selective DNA loss as a general means of gene control, it is necessary to consider the possible role of gene amplification in the regulation of gene expression (for reviews see Stark & Wahl 1984, Kafatos *et al.* 1985). A possible mechanism of gene control (Fig. 2.9) would involve the selective amplification of genes that were expressed at high levels in a particular tissue, the high expression

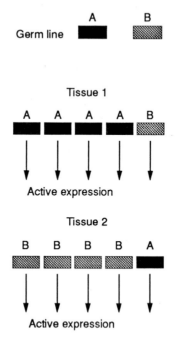

Figure 2.9 Model for gene regulation in which a gene that is expressed at high levels in a particular tissue is selectively amplified in that tissue. Gene A is amplified in tissue 1 and gene B in tissue 2.

24

of such genes simply resulting from normal rates of transcription of the multiple copies of the gene. Unlike gene-deletion models, this possibility is not incompatible with the ability of nuclei from differentiated cells, such as frog intestinal cells, to regenerate a new organism (see Section 2.2.3). Thus, although amplification of intestine-specific genes would have occurred in such cells, this would presumably not prevent the amplification of other tissue-specific genes in the individual cell types of the new organism. The evidence against this idea comes from chromosomal and molecular rather than functional investigations.

2.3.2   Chromosomal studies

As with DNA loss, the identity of the chromosomal complement in individual cell types provides strong evidence against the occurrence of large-scale DNA amplification. None the less, some cases of such amplification have been observed by this means and these will be discussed.

CHROMOSOME POLYTENIZATION IN *DROSOPHILA*

One such case, which has been extensively used in cytogenetic studies, occurs in the salivary gland cells of the fruit fly *Drosophila melanogaster*. In these cells the DNA replicates repeatedly and the daughter molecules do not separate but remain close together, forming a single giant, or polytene, chromosome (Fig. 2.10) which contains approximately 1000 DNA molecules. Although such giant chromosomes have proved very useful in experiments involving visualization of DNA transcription (see Ch. 3, Section 3.2.4), they cannot really be considered an example of gene control since virtually the whole of the genome (with the exception of some repeated DNA) participates in this amplification and there is no evidence for selective amplification of the genes for proteins required in the salivary gland.

DNA AMPLIFICATION AND DELETION IN CILIATED PROTOZOA

A similar polytenization also occurs in many genera of ciliated Protozoa, such as *Oxytrichia* and *Tetrahymena*. In these unicellular organisms development is controlled by separate germ line micronuclei and somatic macronuclei. As with the multicellular organisms discussed in Section 2.2.2, a selective loss of DNA occurs in the somatic nucleus, involving DNA required only in the germ line. Unlike the multicellular organisms, however, this is preceded by a polytenization of the macronuclear DNA, which is followed by the fragmentation and degradation of up to 95 per cent of the DNA (Howard & Blackburn

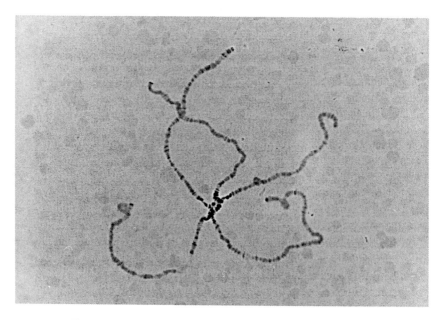

Figure 2.10 Polytene chromosomes of *Drosophila melanogaster*.

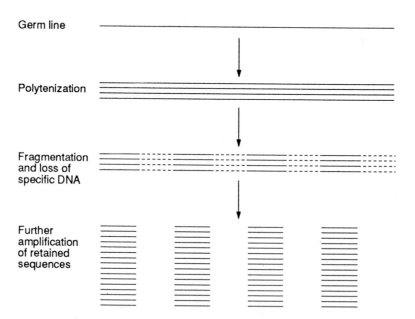

Germ line

Polytenization

Fragmentation and loss of specific DNA

Further amplification of retained sequences

Figure 2.11 Stages in the amplification of the macronuclear DNA of ciliated Protozoa.

1985). Interestingly, the fragments that survive this degradation process and contain the genes required for somatic growth then undergo further rounds of amplification to produce the mature macronuclear DNA (Fig. 2.11). Hence the somatic macronuclear DNA is created by a non-selective amplification followed by selective degradation and further amplification of the DNA. Presumably the many copies of the genes for structural proteins retained in the macronucleus are required to produce the large amounts of these proteins necessary in these very large single-celled organisms.

AMPLIFICATION OF RIBOSOMAL DNA

In addition to the amplification of chromosomal DNA in these ciliated Protozoa, a separate amplification event specifically increases the copy number of the DNA encoding the ribosomal RNA molecules. After fertilization, the single copy of this DNA present in the germ line is excised from the chromosomal DNA as a single 10.5 kb molecule and, after forming an almost perfect dimeric molecule, is subsequently extensively replicated extra-chromosomally (Figure 2.12; Yao *et al.* 1985).

Such amplification of the ribosomal DNA in order to produce the extensive amounts of structural RNAs required by the ribosomes is also observed in the embryonic development of multicellular organisms,

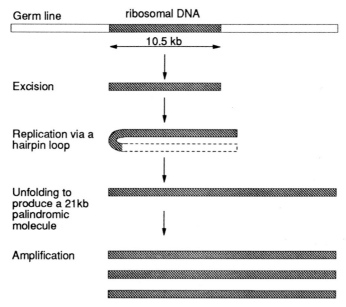

Figure 2.12 Stages in the amplification of the single copy of the DNA encoding the ribosomal RNA in ciliated Protozoa.

even though the germ line DNA of these organisms contains multiple copies of the ribosomal DNA prior to any amplification event. Thus, early in oogenesis the frog oocyte, like all other frog cells, contains approximately 900 chromosomal copies of the DNA encoding ribosomal RNA. During oogenesis, some of the copies are excised from the genome and replicate extensively to produce up to 2 million extra-chromosomal copies of the ribosomal DNA (Brown & Dawid 1968).

These cases of ribosomal DNA amplification clearly represent a response to the intense demand for ribosomal RNA during embryonic development. This requirement has to be met by DNA amplification since the ribosomal RNA represents the final product of ribosomal DNA expression. By contrast, in the case of protein-coding genes, high-level expression can be achieved both by very efficient transcription of the DNA into RNA and also by similarly efficient translation of the RNA into protein. Hence the observed amplification of ribosomal RNA genes does not imply that this mechanism is used in higher organisms to amplify specific protein-coding genes, indeed the general similarity of the chromosome complement in different tissues argues against large-scale amplification of such genes. As with gene deletion, however, the sensitivity of such techniques is such that small-scale amplification of a few genes would not be detected, and it is necessary to use molecular techniques to see whether such amplification occurs.

2.3.3 Molecular studies

In addition to its use in searching for gene loss (see Section 2.2.4) the technique of Southern blotting can also be used to search for DNA amplifications in tissues expressing a particular gene. Thus, either new DNA bands hybridizing to the specific probe will appear, or the same band, present in other tissues, will be observed but it will hybridize more intensely because more copies are present. Despite very many studies of this type only a very small number of cases of such amplification have been reported, the vast majority of tissue-specific genes being present in the same number of copies in all tissues. Thus, for example, the fibroin gene is detectable at similar copy number in the DNA of the posterior silk gland of the silk moth, where it is expressed, and in the DNA of the middle silk gland, where no expression is detectable (Fig. 2.13; Manning & Gage 1978).

However, one exception to this general lack of amplification of protein-coding genes is particular noteworthy. This involves the genes encoding the eggshell, or chorion, proteins in *Drosophila melanogaster*. As first shown by Spradling & Mahowald (1980), the chorion genes are selectively amplified (up to 64 times) in the DNA of the cells which surround the egg follicle, allowing the synthesis in these cells of the

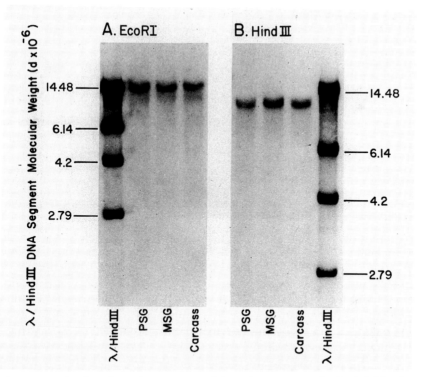

Figure 2.13 Southern blot of DNA prepared from the posterior silk gland (PSG), the middle silk gland (MSG), or the carcass of the silk moth, *Bombyx mori*, with a probe specific for the fibroin gene. Note the identical size and intensity of the band produced by *Eco*RI or *Hind*III digestion in each tissue, although the gene is only expressed in the posterior silk gland.

large amounts of chorion mRNA and protein needed to construct the eggshell. Such amplification can be observed readily in a Southern blot experiment using a recombinant DNA probe derived from one of the chorion genes (Fig. 2.14). Unlike ribosomal DNA, amplification occurs within the chromosome without excision of the template DNA or the newly replicated copies.

This amplification event is of particular interest because it occurs in normal cells as part of normal embryonic development, in contrast, for example, to the amplification of cellular oncogenes that occurs in some cancer cells (see Ch. 8, Section 8.3) or the amplification of the gene encoding the enzyme dihydrofolate reductase (which occurs in mammalian cells in response to exposure to the drug methotrexate; Schimke 1980). None the less, the amplification of the chorion genes appears to be a response to a very specialized set of circumstances, requiring a novel means of gene regulation. Thus eggshell construction in *Drosophila* occurs over a very short period (about 5 h) and probably

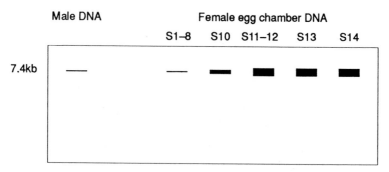

Figure 2.14 Amplification of the chorion genes in the ovarian follicle cells of *Drosophila melanogaster*. Note the dramatic increase in the intensity of hybridization of the chorion-gene-specific probe to the DNA samples prepared from Stage 10 to Stage 14 egg chambers compared to that seen in Stages 1–8 or in male DNA.

necessitates the synthesis of mRNA for the chorion proteins at very high rates, too large to be achieved even by high-level transcription of a single unique gene. It is this combination of high-level synthesis and a very short period to achieve it which necessitates the use of gene amplification. The restricted use that is made of this mechanism is well illustrated, however, by the fact that in silk moths, in which eggshell production occurs by a similar mechanism, multiple copies of the genes for the chorion proteins are present in the germ line. Thus they are present in all somatic cells, including those that do not express these genes, and are not further amplified in the ovarian follicle cells.

Such a finding serves to reinforce the conclusion of both molecular and chromosomal experiments that, as with DNA loss, DNA amplification is not a general method of gene regulation and is used only in isolated specialized cases.

2.4 DNA REARRANGEMENT

Having established that, generally, gene regulation does not occur by DNA deletion or amplification, it is necessary to consider the possibility that it might occur via specific rearrangements of genes in tissues where their protein products are required. Such activation could occur, for example, by translocation of the gene, removing it from a repressive effect of neighbouring sequences, or by bringing together the part of the gene encoding the protein and a promoter element necessary for its transcription (Fig. 2.15).

As with DNA amplification, the occurrence of such rearrangements in one particular cell type might not preclude the subsequent rearrangement of genes required in other tissues. Hence the ability of a

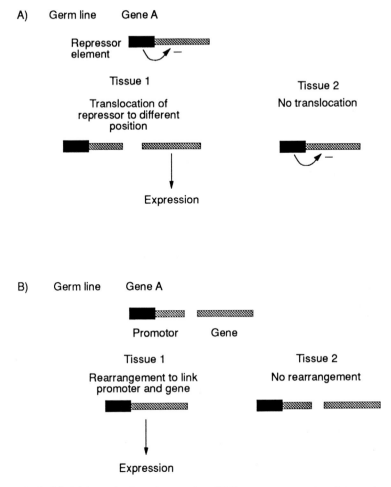

Figure 2.15 Model for activation of a gene by a DNA rearrangement involving either the removal of a repressor element (A) or the joining of the gene and its promoter (B).

single differentiated cell type to give rise to a variety of other differentiated cells in functional studies (see Section 2.2.3.) cannot be used to argue against the occurrence of such rearrangements. Similarly, these rearrangements might be on too small a scale to be detectable in chromosomal studies.

The evidence from molecular studies, however, shows unequivocally that these types of rearrangement are not involved in the selective activation of the vast majority of genes in particular tissues. Thus such rearrangements would produce novel bands when the DNA from a tissue in which the gene was active was cut with restriction enzymes and used in Southern blot experiments (Fig. 2.16). In fact no such

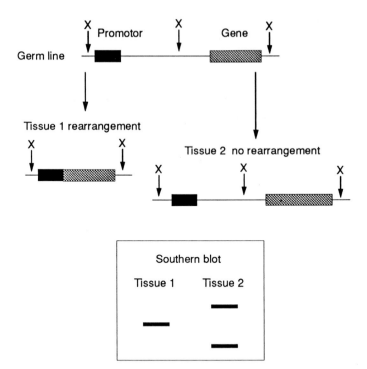

Figure 2.16 Generation of new restriction fragments detectable on a Southern blot following gene rearrangement. Sites for a particular restriction enzyme are indicated by an X and the consequences of digestion with this enzyme of unrearranged or rearranged DNA are indicated.

novel bands are observed in the vast majority of cases when the DNA of a tissue in which the gene is active is compared with tissues where it is inactive (Fig. 2.8 & 2.13).

The evidence against DNA rearrangement as a general means of gene control, obtained in such Southern blot experiments, has been abundantly confirmed by experiments in which the DNA sequence of a particular gene has been obtained both from a tissue where it is inactive and from one where it is active. Such experiments have failed to reveal even a single base difference in these two situations, and indicate that the gene is potentially fully functional in tissues where it is inactive. Despite these clear findings that, as with deletion and amplification, DNA rearrangement is not a general means of gene control, a few cases where it is used have been reported and these will now be considered.

YEAST MATING TYPE

In yeast, mating involves the fusion of two opposite mating types, known as a and α, to generate a diploid cell. In some yeast strains, known as heterothallic strains, these two mating types are entirely separate, as in higher organisms. In homothallic yeasts, by contrast, a cell can switch its mating type from a to α, or vice-versa, and after such a switch it behaves exactly as does any other cell with its new mating type (reviewed by Nasmyth 1982). This switching takes place at a precise time in the life cycle, after a cell has given rise to a daughter by budding (Fig. 2.17). Only the mother cell that produced the daughter undergoes switching, the daughter cell does not switch until it has grown and produced a daughter itself. Repeated cycles of switching from a to α and back occur in this way throughout the life of the organism. Although the reasons for this switching process are unclear, it probably represents a response to the need for rapid fusion of the haploid a and α cells to produce a diploid zygote. The switching of mating type in some of the progeny of a single cell allows this to occur without the requirement for contact with another strain of a different mating type.

A series of elegant genetic experiments showed that whether a cell was a or α in mating type was controlled by whether it possessed an *a* or α gene at the transcriptionally active *MAT* (or mating type locus) on chromosome 3. In addition to this active copy, yeast cells also possess transcriptionally silent copies of both *a* and α genes elsewhere on chromosome 3 (known as the *HML* and *HMR* loci). Switching occurs by a cassette mechanism in which one of these inactive copies replaces the

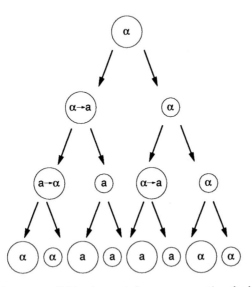

Figure 2.17 Mating type switching in yeast. In every generation the larger mother cell, which has produced a smaller daughter, switches its genotype from *a* to α or vice-versa.

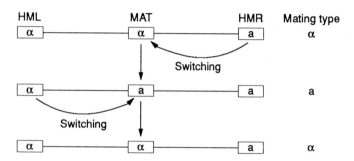

Figure 2.18 Mechanism of mating type switching involving the movement of an *a* or α gene from the inactive *HML* or *HMR* loci to the active *MAT* locus.

active copy in the *MAT* locus. When an *a* gene in the *MAT* locus is replaced by an α gene, the α gene becomes active and the mating type of the cell switches (Fig. 2.18). The reverse switch. in which an α gene at the *MAT* locus is replaced by an *a* gene allows the switch from α to a to occur.

The process of switching is catalysed by an endonuclease, which is the product of the *HO* (or homothallism) gene and which makes a double-stranded cut in the DNA at the *MAT* locus, initiating the transfer process. HO is only synthesized during a short part of the G_1 phase of the cell cycle in mother cells and not at all in daughter cells. Hence, only mother cells and not daughter cells switch. However, such a finding merely leads to the problem of why the *HO* gene should be transcribed only in mother cells.

The regulation of *HO* transcription has been shown to involve a negative regulator known as SIN3, which represses *HO* transcription, and another regulator, SW15, which allows transcription even in the presence of SIN3 (reviewed by Hicks 1987). It has been postulated (Sternberg *et al.* 1987) that both SW15 and SIN3 are expressed at defined times in G_1, with SW15 expressed first. In mother cells SW15 and SIN3 are present together at the only point in the cell cycle when *HO* can be expressed. Thus SW15 antagonizes the repressive action of SIN3, and HO is made (Fig. 2.19). In daughter cells, which are smaller, much more time is spent in G_1 and SW15 levels have decayed away before the window in which *HO* can be expressed is reached. Hence only SIN3 is present at this time and no transcription occurs.

The great attraction of this model is that it does not explain *HO* regulation merely in terms of an event, such as asymmetric partitioning of regulatory molecules between mother and daughter cells, which would in turn require an explanation. Rather, it relies on an entirely independent primary regulatory event, namely the difference in the timing of the cell division cycle between large and small cells. Thus it

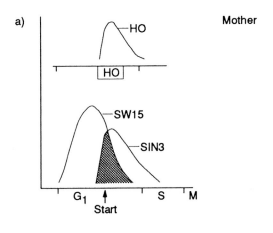

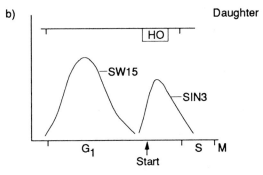

Figure 2.19 Model for the regulation of *HO* gene transcription in mother and daughter cells by the antagonistic action of the SIN3 and SW15 regulators. In mother cells (a) SW15 expression overlaps that of SIN3 and antagonizes it, allowing HO to be made. In daughter cells (b), which are smaller and hence have a longer G_1 period, only SIN3 is present at the critical period for *HO* expression, SIN3 thus represses HO synthesis and switching does not occur.

provides a mechanism for initiating the regulatory process rather than merely setting the question back one step.

This process, worked out by the use of the powerful genetic techniques available in yeast, may serve as a model for the regulation of developmental processes in higher organisms. Indeed, Herskowitz (1985) has drawn the analogy between the switching of mating type in yeast and the creation of a stem cell lineage in the development of higher organisms. Thus, if the diagram of switching shown in Figure 2.17 is modified, by assuming first that switching is irreversible and secondly by considering that α represents a stem cell while a represents a differentiated derivative, the model in Figure 2.20 is obtained. In this model lineage, a stem cell is continually dividing, generating one daughter which differentiates while the other maintains

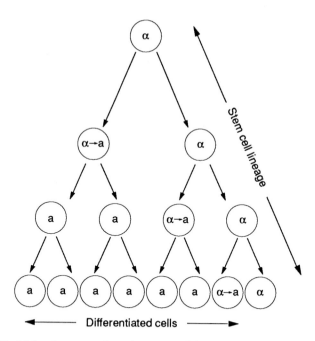

Figure 2.20 Model for the generation of a stem cell lineage, producing differentiated cells by a system based on α to a mating type switching. In this model, α represents the stem cell and a the differentiated cell, and it is assumed that, unlike the yeast mating type system, the α to a switch is irreversible.

the stem cell lineage. This is a very common feature in the development of higher organisms and has already been discussed in the erythrocyte lineage earlier in this chapter (see Section 2.2.2).

Even if this model for the generation of a stem cell lineage is correct, it does not imply that the switch process need take place via a DNA rearrangement rather than by any other control mechanism that would allow the cell to change from one state to another. However, the possibility that such switches do occur in the development of higher organisms, by whatever means, needs to be considered, and the mating type system offers a model for how this might be regulated.

Interestingly, the mating type system also involves other mechanisms of gene regulation that may be relevant to higher organisms. Thus the silent copies of the *a* and α genes in the HML and HMR loci are maintained in a transcriptionally inactive form within a tightly packed chromatin structure, a process which is also involved in the regulation of gene expression in mammals and other higher eukaryotes (see Chapter 5, Section 5.4). Similarly, the a and α gene products produce, respectively, the a and α phenotypes by activating the transcription of sets of other genes (the a and α specific genes). Most interestingly, the structure of the a and α gene products bears similarities to the homeobox proteins, which are believed to play a role in the activation of gene transcription during the development of higher organisms such as mammals and *Drosophila* (see Ch. 7, Section 7.2.2).

It is clear, therefore, that the study of the yeast mating type system, which is very amenable to genetic analysis, can offer insights into the mechanisms being used to regulate gene expression in higher organisms. It is unlikely, however, that the use of DNA rearrangement to regulate the expression of the *a* and α control loci in this system indicates that this type of mechanism is used widely to control the expression of regulatory proteins in higher organisms. Rather, it appears to represent a response to the need to express *a* or α but not both, and to continually switch between the a and α states.

ANTIGENIC VARIATION IN TRYPANOSOMES

Clearly, a similar switch mechanism to that used in yeast would be ideal in a situation where it was necessary to express successively one of a series of particular genes. In this situation one of many different genes present in inactive reservoir sites could be brought to the active site and expressed, only to be subsequently replaced by another gene from the reservoir. This mechanism is indeed used for precisely this purpose to control antigenic variation in trypanosomes (reviewed by van der Ploeg 1987). These unicellular Protozoa avoid the defensive response mounted by the immune system of the mammals they parasitize by continually changing the structure of their outer surface glycoprotein, which is accordingly known as the variant surface glycoprotein (VSG). This is achieved by repeatedly replacing the copy of the VSG gene at an expression site with one of approximately 1000 different VSG genes present at other sites in the genome (Fig. 2.21). At the expression site the VSG gene becomes linked to sequences required for its transcription and hence is actively transcribed. The copies present at other sites lack these sequences and hence they are inactive. The differences in the protein encoded by the different VSG genes that are successively placed in the active site thus produce the observed variation in the surface glycoprotein that occurs during the progress of an infection and prevents the mounting of an effective immune response by the parasitized host.

ANTIBODY PRODUCTION IN MAMMALIAN B-CELLS

When mammals are exposed to foreign bacteria or viruses, their immune systems respond by synthesizing specific antibodies directed against the proteins of these organisms, with the aim of neutralizing the harmful effects of the infection. The potential requirement for the synthesis of diverse kinds of antibody by the mammalian immune

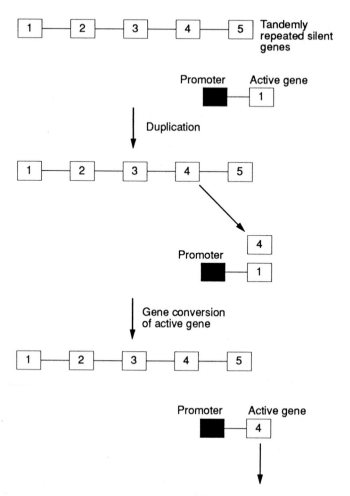

Figure 2.21 Switching of surface antigen expression in trypanosomes is achieved by the duplication of one of a tandem array of silent genes and its replacement of the active gene next to the promoter by a process of gene conversion.

system far exceeds the diversity of different VSG proteins encoded by the trypanosome genome discussed in the preceding section. Thus, not only must the body be capable of producing different highly specific antibodies against all the different VSG proteins, but it must also be able to defend itself against a bewildering variety of challenges from other infectious organisms by the production of similarly specific antibodies against their proteins (for a recent review of the immune system, emphasizing the molecular aspects, see Watson *et al.* 1987). These antibodies, or immunoglobulins, are produced by the covalent association of two identical heavy chains and two identical light chains

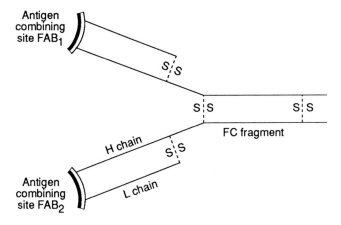

Figure 2.22 Association of two heavy (H) and two light (L) chains to form a functional antibody molecule. The disulphide bonds linking the chains and the region that binds antigen are indicated.

to produce a functional molecule (Fig. 2.22). The combination of specific heavy and light chains produces the specificity of the antibody molecule. In particular, each chain contains, in addition to a constant region (which is relatively similar in different antibody molecules), a highly variable region which differs widely in amino-acid sequence in different antibodies. It is this variable region that actually interacts with the antigen and hence determines the specificity of the antibody molecule.

As any heavy chain can associate with any light chain, and as approximately 1 million types of antibody specificities can be produced, at least 1000 genes encoding different heavy chains and a similar number encoding the light chains would be required. Copy-number studies have shown, however, that the germ line does not contain anything like this number of intact immunoglobulin genes, and no specific amplification events have been detected in the DNA of antibody-producing B-cells.

Rather, as first suggested by Dreyer & Bennett (1965), functional immunoglobulin genes are created by DNA rearrangements in the B-cell lineage (reviewed by Tonegawa 1983). Thus the germ-line DNA contains a large number of tandemly repeated DNA segments encoding different variable regions, which are separated by over 100 kbp from a much smaller number of DNA segments encoding the constant region of the molecule (Fig. 2.23). This organization is maintained in most somatic cell types, but in each individual B-cell a unique DNA rearrangement event occurs by which one specific variable region is brought together with one constant region by deletion of the intervening DNA. In this manner, a different functional

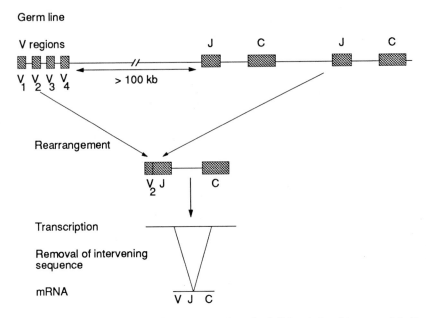

Figure 2.23 Rearrangement of the gene encoding the λ light chain of immunoglobulin results in the linkage of one specific variable region (V) to a specific joining region (J) with removal of the intervening DNA. The gene is then transcribed into RNA and the region between the joining (J) and constant regions (C) is removed by RNA splicing to create a functional mRNA.

gene, containing specific variable and constant regions, is produced in each individual B-cell.

These rearrangements can be detected readily by the appearance of novel bands in Southern blot analyses of the structure of the immunoglobulin genes in the DNA of antibody-producing B-cells. In the example shown in Figure 2.24 (based on the data of Hozumi & Tonegawa 1976) the digestion of DNA from non-B-cells with the restriction enzyme *Bam*H1 produces fragments of 6.5 kb and 9.5 kb containing the variable and constant regions, respectively; whereas digestion of rearranged B-cell DNA generates a single 3.5 kb band containing both parts of the gene. This combination of different variable (V) and constant (C) regions in the production of a functional immunoglobulin gene obviously creates a wide range of different potential antibody specificities by the joining of particular V and C segments.

Despite this, more and more complex rearrangements are used in different types of immunoglobulin gene to generate still more potential variety. Thus, in the simplest case that we have discussed so far, that of the λ light chain (Fig. 2.23), one of many V regions is brought into

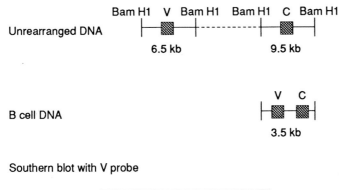

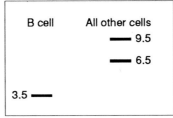

Southern blot with V probe

Figure 2.24 The rearrangement of the immunoglobulin gene locus creates a different *Bam*H1 restriction fragment hybridizing to immunoglobulin DNA-specific probes in the DNA of B-cells, compared to those that are present in other cell types where no rearrangement has occurred.

association with a constant region, consisting of a single J segment (encoding the junction between the variable and constant regions) which is separated by a short region from the DNA encoding the C segment itself. The rearranged gene is then transcribed and the intervening sequence between the J and C segments removed by RNA splicing to produce a functional messenger RNA (see Ch. 4, Section 4.2). In the genes for another type of light chain, known as κ, the situation is more complex. Thus these genes contain multiple J segments linked to individual constant segments and hence further diversity is created by the recombination of one of many V regions with one of several different J regions.

The case of the genes encoding the heavy-chain molecule is still more complex (Fig. 2.25) in that these genes also contain a D, or diversity, segment (Early *et al.* 1980). A functional heavy-chain gene is thus created by two successive rearrangements. In the first of these one of several D segments is brought together with one of the J segments and the associated C region. Subsequently, a V region segment is recombined with this DJC element exactly as occurs in the VJ joining found in the light-chain genes. These two combinatorial events allow the selection of one particular copy of each of four regions to

41

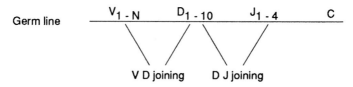

Figure 2.25 Rearrangement in the gene encoding the heavy chain of the immunoglobulin molecule involves joining of a variable (V) and a diversity (D) region as well as the similar joining of a diversity and a junction (J) region, with removal of the intervening DNA in each case.

produce the functional gene, generating many different possible heavy-chain molecules.

Interestingly, the generation of diversity by DNA rearrangement is supplemented by another DNA alteration, so far unique in mammalian cells, to the genes of the immune system. It appears that immunoglobulin genes can undergo point mutations within particular B-cells, resulting in changes in the amino acids they encode and hence in their specificity. Such somatic mutation (reviewed by Baltimore 1981) produces forms of immunoglobulin not encoded by the DNA present in other tissues or in the germ line, and represents another form of tissue-specific gene alteration occurring within the immunoglobulin genes.

It is clear that the primary role of the rearrangements and alterations which occur in the immunoglobulin genes in B-cells is the generation of the diversity of antibodies required to cope with the vast number of different possible antigens, rather than to produce the high-level expression of the immunoglobulin genes which occurs in such cells. None the less, high-level immunoglobulin-gene expression does occur as a consequence of such rearrangements. Thus the DNA encoding the variable regions of the antibody molecule is closely linked to the promoter elements that direct the transcription of the gene. Such elements show no activity in the unrearranged DNA of non-B-cells, but become fully active only when the variable region is joined to the constant region. In the case of the heavy-chain genes (Banerji *et al.* 1983, Gilles *et al.* 1983), this is because the intervening sequence between the J and C regions of these genes contains an enhancer element (see Ch. 6, Section 6.3) which, although not a promoter itself, greatly increases the activity of the promoter adjacent to the V-region element. Hence the rearrangement not only brings the promoter element into a position where it can produce a functional mRNA containing VDJ and C, regions but also facilitates the high-level transcription of this gene by juxtaposing promoter and enhancer elements that are separated by over 100 kb of DNA before the rearrangement event (Fig. 2.26).

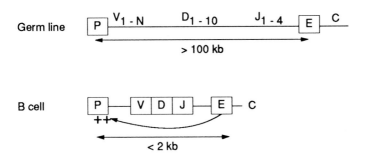

Figure 2.26 The rearrangement of the heavy-chain gene brings an enhancer (E) in the intervening sequence between J and C close to the promoter (P) adjacent to V, and results in the activation of the promoter and gene transcription.

The element of gene control in this process is, however, entirely secondary to the need to produce diversity by rearrangement. Thus, if the immunoglobulin enhancer is linked artificially close to the immunoglobulin promoter and introduced into a variety of cells, it will activate transcription from the promoter only in B-cells and not in other cell types. Hence the enhancer is active only in B-cells and would not activate immunoglobulin transcription even if functional immuno-globulin genes containing closely linked VDJ and C segments were present in all other tissues. The immunoglobulin genes are therefore regulated by tissue-specific activator sequences in much the same way as other genes (see Ch. 6), but such activation has been made to depend on the occurrence of a DNA rearrangement whose primary role lies elsewhere.

Thus, although similar DNA rearrangements are used in other situa-tions where diversity must be generated, for example, in the genes encoding the T-cell receptor which is involved in the response to foreign antigen of T-lymphocytes (reviewed by Davis 1985), it is clear that they do not represent a general means of gene regulation applicable to other situations. Such a conclusion is entirely in accordance with the results of Southern blot and sequencing experi-ments which, as previously discussed, have failed to detect such rearrangements in the vast majority of genes active only in particular tissues.

2.5   CONCLUSIONS

Although we have discussed in this chapter a number of cases where DNA loss, amplification, or rearrangement regulate gene expression, it is clear that these represent isolated special cases where the dictates of a particular situation have necessitated the use of such mechanisms.

Thus the loss of DNA in red blood cells is dictated by the need to fill the cell with haemoglobin, the amplification of the chorion genes in *Drosophila* by the requirement to produce the corresponding proteins in a very short time; and the rearrangement of the immunoglobulin genes by the need to generate a diverse array of antibodies. Hence these cases are not representative of a general mechanism of gene regulation by DNA alteration and, in general, the DNA of different cell types is quantitatively and qualitatively identical. Given that we have established previously that the RNA content of different cell types can vary dramatically, it is now necessary to investigate how such differences in RNA content can be produced from the similar DNA present in such cell types.

REFERENCES

Baltimore, D. 1981. Somatic mutation gains its place among the generators of diversity. *Cell* **26**, 295–6.

Banerji, J., L. Olson & W. Schaffner 1983. A lymphocyte-specific cellular enhancer is located downstream of the joining region in immunoglobulin heavy chain genes. *Cell* **33**, 729–40.

Bessis, N. & M. Brickal 1952. Aspect dynamique des cellules du sang. *Revues Hématologique* **7**, 407–35.

Brown, D. D. 1981. Gene expression in eukaryotes. *Science* **211**, 667–74.

Brown, D. D. & B. Dawid 1968. Specific gene amplification in oocytes. *Science* **160**, 272–80.

Davis, M. 1985. Molecular genetics of the T cell receptor beta chain. *Annual Review of Immunology* **3**, 537–60.

Dreyer, W. J. & J. C. Bennett 1965. The molecular basis of antibody formation: a paradox. *Proceedings of the National Academy of Sciences of the USA* **54**, 864–9.

Early, P., H. Huang, M. Davis, K. Calame, & L. Hood 1980. An immunoglobulin heavy chain variable region is generated from three segments of DNA: V_H, D and J_H. *Cell* **19**, 981–92.

Gillies, S. D., S. L. Morrison, V. T. Oi & S. Tonegawa 1983. A tissue specific enhancer is located downstream of the joining region in immunoglobulin heavy chain genes. *Cell* **33**, 717–28.

Gurdon, J. B. 1968. Transplanted nuclei and cell differentiation. *Scientific American* **219**, (Dec.), 24–35.

Herskowitz, L. 1985. Master regulatory loci in yeast and lambda. *Cold Spring Harbor Symposia* **50**, 565–74.

Hicks, J. B. 1987. Mechanisms of differentiation. *Nature* **326**, 444–5.

Howard, E. A. & E. H. Blackburn 1985. Reproducible and variable genomic rearrangements occur in the developing somatic nucleus of the ciliate *Tetrahymena thermophila*. *Molecular and Cellular Biology* **5**, 2039–50.

Hozumi, N. & S. Tonegawa 1976. Evidence for somatic rearrangement of immunoglobulin genes coding for variable and constant regions. *Proceedings of the National Academy of Sciences of the USA* **73**, 3628–32.

Jefferys, A. J. & R. A. Flavell 1977. The rabbit β-globin gene contains a large insert in the coding sequence. *Cell* **12**, 1097–108.

John, B. & G. L. G. Miklos 1979. Functional aspects of satellite DNA and heterochromatin. *International Review of Cytology* **58**, 1–113.

Kafatos, F. C., W. Orr & C. Delidakis 1985. Developmentally regulated gene amplification. *Trends in Genetics* **1**, 301–6.

Manning, R. F. & L. P. Gage 1978. Physical map of the *Bombyx mori* DNA containing the gene for silk fibroin. *Journal of Biological Chemistry* **253**, 2044–52.

Nasmyth, K. A. 1982. Molecular genetics of yeast mating type. *Annual Review of Genetics* **16**, 439–500.

Ploeg, L. H. T. van der. 1987. Control of variant antigen switching in Trypanosomes. *Cell* **51**, 159–61.

Schimke, R. T. 1980. Gene amplification and drug resistance. *Scientific American* **243**, (Nov.) 50–9.

Southern, E. M. 1975. Detection of specific sequences among DNA fragments separated by gel electrophoresis. *Journal of Molecular Biology* **98**, 503–17.

Spradling, A. C. & A. P. Mahowald 1980. Amplification of genes for chorion proteins during oogenesis in *Drosophila melanogaster*. *Proceedings of the National Academy of Sciences of the USA* **77**, 1096–100.

Stark, G. R. & G. M. Wahl 1984. Gene amplification. *Annual Review of Biochemistry* **53**, 447–91.

Sternberg, P. W., M. J. Stern, I. Clark & I. Herskowitz 1987. Activation of the yeast *HO* gene by release from multiple negative controls. *Cell* **48**, 567–77.

Steward, F. C. 1970. From cultured cells to whole plants: the induction and control of their growth and morphogenesis. *Proceedings of the Royal Society, Series B* **175**, 1–30.

Tonegawa, S. 1983. Somatic generation of antibody diversity. *Nature* **302**, 575–81.

Watson, J. D., N. H. Hopkins, J. W. Roberts, J. A. Steitz & A. M. Weiner 1987. *Molecular Biology of the gene*, 4th edn, Vol. 2, 832–97. Menlo Park, California: Benjamin/Cummings.

Wilson, E. B. 1928. *The cell in development and heredity*. New York: Macmillan.

Yamada, T. 1967. Cellular and sub-cellular events in Wolffian lens regeneration. *Current Topics in Developmental Biology* **2**, 249–83.

Yao, M.-C., S.-G. Zhu & C.-H. Yao 1985. Gene amplification in *Tetrahymena thermophila*: formation of extra-chromosomal palindromic genes coding for rRNA. *Molecular and Cellular Biology* **5**, 1260–7.

CHAPTER THREE

Regulation at transcription

3.1 LEVELS OF GENE REGULATION

The observation that differences in the RNA and protein content of different tissues are not paralleled by significant differences in their DNA content indicates that the process whereby DNA produces messenger RNA must be the level at which gene expression is regulated in eukaryotes.

In bacteria this process involves only a single stage, that of transcription, in which an RNA copy of the DNA is produced by the enzyme RNA polymerase. Even while this process is still occurring, ribosomes attach to the nascent RNA chain and begin to translate it into protein. Hence cases of gene regulation in bacteria, such as the switching on of the synthesis of the enzyme β-galactosidase in response to the presence of lactose (its substrate), are mediated by increased transcription of the appropriate gene (Jacob & Monod 1961).

Clearly, a similar regulation of gene transcription in different tissues, or in response to substances such as steroid hormones which induce the synthesis of new proteins, represents an attractive method of gene regulation in eukaryotes. In contrast to the situation in bacteria, however, a number of stages intervene between the initial synthesis of the primary RNA transcript and the eventual production of messenger RNA (for a review see Nevins 1983) (Fig. 3.1). The initial transcript is modified at its 5' end, by the addition of a cap structure containing a modified guanosine residue, and is subsequently cleaved near its 3' end, followed by the addition of up to 200 adenosine residues in a process known as polyadenylation. Subsequently, intervening sequences, or introns, which interrupt the protein coding sequence in both the DNA and the primary transcript of many genes (Jeffreys & Flavell 1977, Tilghman *et al.* 1978) are removed by a process of RNA splicing (for a review see Sharp 1987). Although this produces a functional messenger RNA, the spliced molecule must then be

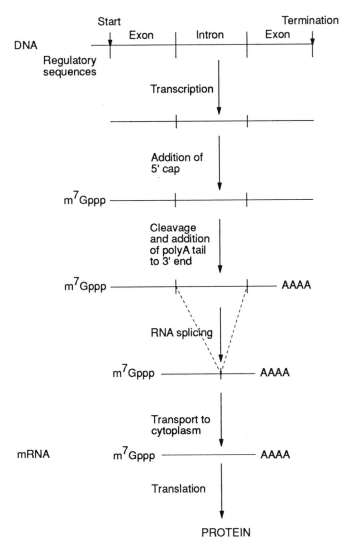

Figure 3.1 Stages in eukaryotic gene expression which could be regulated.

transported from the nucleus, where these processes occur, to the cytoplasm where it can be translated into protein.

In theory any one of these stages, all of which are essential for the production of a functional mRNA, could be used to regulate the expression of specific genes in particular tissues. For example, it has been suggested that gene regulation could occur by transcribing all genes in all tissues and selecting which transcripts were appropriately spliced by correct removal of intervening sequences thereby producing a functional mRNA (Davidson & Britten 1979).

The existence of such a plethora of possible regulatory stages has led, therefore, to much investigation as to whether all or any of these are used. In general, such studies have shown that in higher eukaryotes, as in bacteria, the primary control of gene expression is at the level of transcription, and the evidence showing that this is the case is discussed in this chapter. A number of cases of post-transcriptional regulation do exist, however, and these are discussed in Chapter 4.

3.2   EVIDENCE FOR TRANSCRIPTIONAL REGULATION

The evidence for the regulation of gene transcription comes from several types of study, which will be considered in turn.

3.2.1   Evidence from studies of nuclear RNA

If regulation of gene expression takes place at the level of transcription, the differences in the cytoplasmic levels of particular mRNAs which occur between different tissues should be paralleled by similar differences in the levels of these RNAs within the nuclei of different tissues. In contrast, regulatory processes in which a gene was transcribed in all tissues and the resulting transcript either spliced or transported to the cytoplasm in a minority of tissues would result in cases where differences in mRNA content occurred without any corresponding difference in the nuclear RNA (Fig. 3.2). Hence a study of the level of particular RNA species in the nuclear RNA of individual

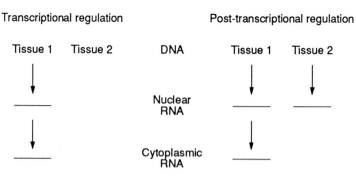

Figure 3.2 Consequences of transcriptional or post-transcriptional regulation on the level of a specific nuclear RNA in a tissue that expresses the corresponding cytoplasmic mRNA (tissue 1) and one that does not (tissue 2).

tissues or cell types serves as an initial test to distinguish transcriptional and post-transcriptional regulation.

The earliest studies in this area focused on the highly abundant RNA species produced during terminal differentiation, which could be readily studied simply because of their abundance. Thus even before the discovery of intervening sequences revealed the possibility of regulation at the level of RNA splicing, Gilmour et al. (1974) studied the processes regulating the accumulation of globin which occur when Friend erythroleukaemia cells are treated with an inducer of globin production, dimethyl sulphoxide. In this system, the accumulation of cytoplasmic globin RNA, which produces the increase in synthesis of globin protein, was paralleled by increasing accumulation of globin RNA within the nucleus, suggesting that the primary effect of the inducer was at the level of globin gene transcription. Similarly, in the experiments of Groudine et al. (1974) globin-specific RNA was readily detectable in the nuclei of globin-synthesizing erythroblasts but not in the nuclear RNA of fibroblasts or muscle cells, hence indicating the involvement of transcriptional control in regulating the abundant production of globin RNA only in the red blood cell lineage and not in other cell types.

Subsequently, the use of Northern blotting (see Ch. 1, Section 1.3.2) allowed the separation of different species within nuclear RNA by size, and their visualization by hybridization to an appropriate probe. In the case of genes with many intervening sequences, such as that encoding the egg protein ovalbumin, many different RNA species could be observed in the nucleus, including not only the primary transcript and the fully spliced RNA prior to transport to the cytoplasm but also a series of intermediate-sized RNAs from which some of the intervening sequences had still to be removed (Roop et al. 1978; Fig. 3.3). The identification of such potential precursors of the mature mRNA allowed a study of their expression in tissues either producing or not producing ovalbumin mRNA and protein. Thus cytoplasmic ovalbumin mRNA and protein are present only in the oviduct following stimulation with oestrogen, and disappear when oestrogen is withdrawn. Similarly, the mRNA and protein are absent in other tissues, such as the liver, and cannot be induced by treatment with the hormone in these tissues. Studies of the distribution of both the fully spliced nuclear RNA and the larger precursors (Roop et al. 1978) showed that these species could only be detected in the nuclear RNA of the oviduct following oestrogen stimulation and were absent in the liver nuclear RNA, or in oviduct nuclear RNA following oestrogen withdrawal (Fig. 3.4). Hence the distribution of these precursors in the nucleus exactly parallels that of the cytoplasmic mRNA, a finding entirely consistent with the transcriptional induction of the ovalbumin

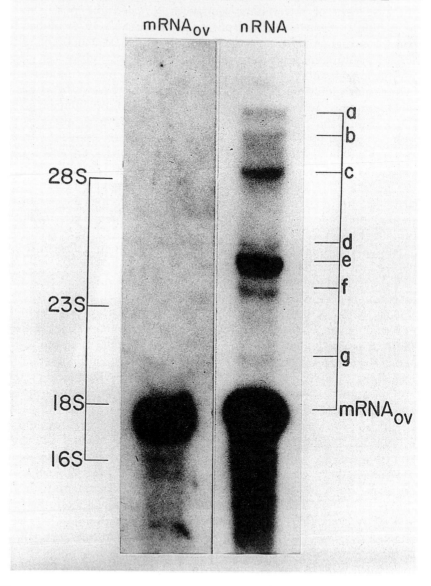

Figure 3.3 Northern blot showing that unspliced and partially spliced precursors (a–g) to the ovalbumin mRNA (mRNA_ov) are detectable in the nuclear RNA (nRNA) of oestrogen-stimulated oviduct tissue.

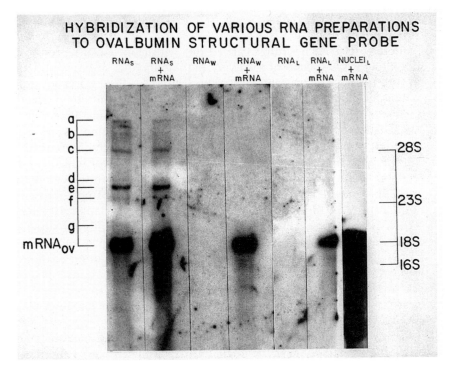

HYBRIDIZATION OF VARIOUS RNA PREPARATIONS TO OVALBUMIN STRUCTURAL GENE PROBE

Figure 3.4 Northern blot showing that the nuclear precursors to ovalbumin mRNA (seen in Fig. 3.3) are detectable in the nuclear RNA of oestrogen-stimulated oviduct (RNA_s) but are absent in the nuclear RNA of oestrogen-withdrawn oviduct (RNA_w) and of liver (RNA_L). Moreover, ovalbumin mRNA mixed with withdrawn oviduct (RNA_w + mRNA) or liver (RNA_L + mRNA) nuclear RNA is not degraded, showing that the absence of RNA for ovalbumin in these nuclear RNAs is not due to a nuclease specifically degrading the ovalbumin RNA.

gene in the oviduct in response to oestrogen. These observations are difficult to reconcile, however, with a model in which the hormone acts to relieve a block in RNA splicing or transport, which exists in untreated oviduct and other tissues, since such models would predict an accumulation of unspliced or untransported RNA for ovalbumin within the nucleus.

These early studies have now been abundantly supplemented by many others measuring the nuclear RNA levels of other specific genes whose expression changes in particular situations, such as those encoding a number of highly abundant mRNAs present only in the soya bean embryo (Goldberg et al. 1981) or that encoding the developmentally regulated mammalian liver protein, α-foetoprotein (Latchman et al. 1984). In general, such studies have led to the conclusion that in most cases alterations in specific mRNA levels in the cytoplasm are accompanied by parallel changes in the levels of the corresponding nuclear RNA species.

Interestingly, such studies carried out with cloned DNA probes for individual RNA species of relatively high abundance are in contrast to reports using R_0t curve analysis (see Ch. 1, Section 1.3.1) to study variations in the total nuclear RNA population in different tissues. Such experiments, which examine mainly low-abundance RNAs, have suggested that in some organisms, such as sea urchins (Wold *et al.* 1978) and tobacco plants (Kamaly & Goldberg 1980), the nuclear RNA is a highly complex mixture of different RNAs, some of which can be retained in the nucleus in one tissue and transported to the cytoplasm in other tissues. These studies, which are indicative of post-transcriptional control in these organisms, are discussed in Chapter 4 (Section 4.1). It is noteworthy, however, that in mammals a different situation exists. Thus in these organisms the increased number of different sequences in nuclear compared to cytoplasmic RNA can be accounted for by the presence of intervening sequences, which are transcribed but not transported to the cytoplasm, without the need to postulate the existence of whole transcripts which are confined to the nucleus in specific tissues.

Hence it appears from these studies that, at least in mammals and other higher eukaryotes, the regulation of transcription, resulting in parallel changes in nuclear and cytoplasmic RNA levels, is the primary means of regulating gene expression. However, the studies described so far suffer from the defect that they measure only steady-state levels of specific nuclear RNAs. It could be argued that the gene is being transcribed in the non-expressing tissue but that the transcript is degraded within the nucleus at such a rate that it cannot be detected in assays of steady-state RNA levels (Fig. 3.5). This degradation in the non-expressing tissue but not in the expressing tissue could be

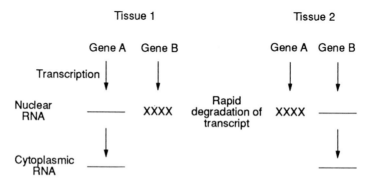

Figure 3.5 Model in which lack of expression of genes in particular tissues is caused by rapid degradation of their RNA transcripts.

produced directly by specific regulation of the rate of turnover of a particular RNA in the different tissues, producing rapid degradation only in the non-expressing tissue. Alternatively, it might result from a block to splicing or transport of the RNA in the non-expressing tissue, followed by the rapid degradation of this unspliced or untransported RNA, which in the expressing tissue would be processed or transported before it could be degraded. These considerations necessitate the direct measurement of gene transcription itself in the different tissues in order to establish unequivocally the existence of transcriptional regulation. Methods to do this have been devised and these will now be discussed.

3.2.2 Evidence from pulse-labelling studies

The synthesis of RNA from DNA by the enzyme RNA polymerase involves the incorporation of ribonucleotides into an RNA chain. Therefore the synthesis of any particular RNA can be measured by adding a radioactive ribonucleotide (usually uridine labelled with tritium) to the cells and measuring how much radioactivity is incorporated into RNA specific for the gene of interest. Clearly, if the degradation mechanisms discussed in the last section do exist, they will, given time, degrade the radioactive RNA molecule produced in this way and drastically reduce the amount of labelled RNA detected. In order to prevent this, the rate of transcription is measured by exposing cells briefly to the labelled uridine in a process referred to as pulse labelling. The labelled uridine is incorporated into nascent RNA chains that are being made at this time and, even before a complete transcript has had time to form, the cells are lysed and total RNA is isolated from them. This RNA, which contains labelled partial transcripts from genes active in the tissue used, is then hybridized to a piece of DNA derived from the gene of interest, the number of radioactive counts that bind providing a measure of the incorporation of labelled precursor into its corresponding RNA (Fig. 3.6).

This method provides the most direct means of measuring transcription and has been used, for example, to show that the induction of globin production which occurs in Friend erythroleukaemia cells in response to treatment with dimethyl sulphoxide is mediated by increased transcription of the globin gene (Lowenhaupt *et al.* 1978). This increased transcription produces the increased levels of globin-specific RNA in the nucleus and cytoplasm of the treated cells, which was discussed earlier in this chapter (see Section 3.2.1).

Although pulse labelling provides a very direct measure of transcription rates, the requirement to use very short labelling times to minimize any effects of RNA degradation limits its applicability. Thus

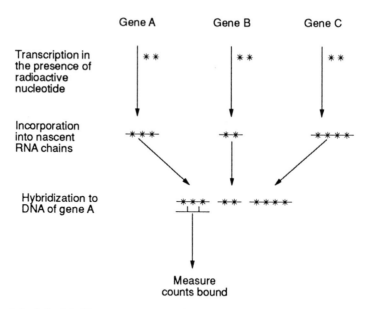

Figure 3.6 Pulse-labelling assay to assess the transcription rate of a specific gene (gene A) by measuring the amount of radioactivity (stars) incorporated into nascent transcripts.

in the experiments of Lowenhaupt *et al.* (1978) it was possible to measure the amount of radioactivity incorporated into globin RNA in the very brief labelling times used (5 or 10 min) only because of the enormous abundance of globin RNA and the very high rate of transcription of the globin gene. With other RNA species, the rates of transcription are insufficient to provide measurable incorporation of label in the short pulse time. More label will, of course, be incorporated if longer pulse times are used, but such pulse times allow the possibility of RNA turnover and are therefore subject to the same objections as the measurement of stable RNA levels.

Hence, although pulse labelling can be used to establish unequivocally that transcriptional control is responsible for the massive synthesis of the highly abundant RNA species present in terminally differentiated cells, it cannot be used to demonstrate the generality of transcriptional control processes and, in particular, their applicability to RNAs which, although regulated in different tissues, never become highly abundant.

This limitation of the pulse-labelling method is especially relevant in view of the existence of the Davidson and Britten model for gene regulation (Davidson & Britten 1979) which specifically postulates that highly abundant RNA species will be regulated in a different manner to the bulk of RNA species, which are of moderate or low abundance. Thus, in this model (Fig. 3.7) all genes are postulated to be transcribed

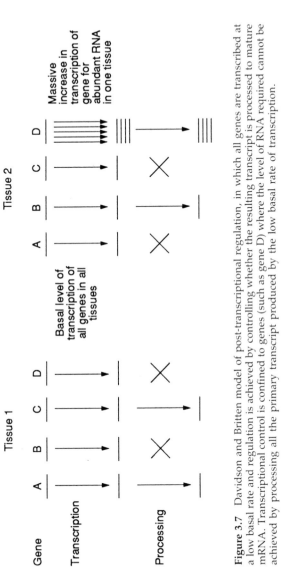

Figure 3.7 Davidson and Britten model of post-transcriptional regulation, in which all genes are transcribed at a low basal rate and regulation is achieved by controlling whether the resulting transcript is processed to mature mRNA. Transcriptional control is confined to genes (such as gene D) where the level of RNA required cannot be achieved by processing all the primary transcript produced by the low basal rate of transcription.

in all tissues at a low basal rate and regulation takes place at a post-transcriptional level by deciding which transcripts are processed to functional mRNA and which are degraded. In the case of most genes the level of RNA and protein required in any particular tissue would be met by processing correctly all of the primary transcript produced by this low basal rate of transcription. The level of the abundant RNA species would be such, however, that they could not be produced with this low rate of gene transcription, even by correctly processing all of the primary transcript. Hence, for the genes encoding these RNA species, a special mechanism would operate and their transcription would be dramatically increased in some tissues, as observed by pulse labelling. This theory postulates that the regulation of gene expression by changes in transcription is confined to the few genes encoding highly abundant RNA species, and that post-transcriptional control processes will regulate the expression of less abundant RNAs whose genes will be transcribed even in tissues where no mRNA is synthesized. In order to test this theory it is necessary to use a method of measuring transcription which, although less direct than pulse labelling, is more sensitive and hence can be applied to a wider variety of cases, including non highly abundant mRNAs. This method is discussed in the next section.

3.2.3 Evidence from nuclear run-on assays

The primary limitation on the sensitivity of pulse labelling is the existence within the cell of a large pool of non-radioactive ribo-nucleotides, which are normally used by the cell to synthesize RNA. When labelled ribonucleotide is added to the cell, it is considerably diluted in this pool of unlabelled precursor. The amount of label incorporated into RNA in the labelling period is therefore very small, since the majority of ribonucleotides incorporated are unlabelled. The sensitivity of this method is thus severely reduced, resulting in its observed applicability only to genes with very high rates of transcription. Interestingly, however, although transcription takes place in the nucleus, most of the pool of precursor ribonucleotides is present in the cytoplasm. Hence it is possible, by removing the cytoplasm and isolating nuclei, to remove much of the pool of unlabelled ribonucleotide. Labelled ribonucleotides can then be added directly to the isolated nuclei in a test-tube and their rate of incorporation into RNA used as a measure of the rate of transcription. Because the labelled ribonucleotides are not diluted in the unlabelled cytoplasmic pool, considerably more label is incorporated into any particular gene than is observed in pulse-labelling experiments. The label incorporated into any particular transcript is detected by hybridization to its corresponding DNA exactly as in pulse-labelling experiments.

This method, which is known as a nuclear run-on assay, is therefore much more widely applicable than pulse labelling, and can be used to quantify the transcription of genes that are never transcribed at levels detectable by pulse labelling. Moreover, very many studies (for a review see Darnell 1982 and references therein) have now established that the RNA synthesized by isolated nuclei in the test-tube is similar to that made by intact whole cells, and that the method is therefore not only sensitive but also provides an accurate measure of transcription, free from artefact.

In initial studies, nuclear run-on assays were used to measure the transcription of highly abundant RNA species. Thus, for example, nuclei isolated from the erythrocytes of adult chickens (which, unlike the equivalent cells in mammals, do not lose their nuclei) were shown to transcribe only the gene encoding the adult β-globin protein, whereas nuclei prepared from similar cells isolated from embryonic chickens failed to transcribe this gene and instead transcribed the gene encoding the form of β-globin made in the embryo (Groudine et al. 1981). Such a finding parallels the observation discussed earlier that the RNA specific for the adult form of β-globin is present only in adult and not in embryonic erythroid cell nuclei (see Section 3.2.1) and indicates that the developmentally regulated production of different forms of globin protein is under transcriptional control.

A similar parallelism between the results of nuclear RNA studies and nuclear run-on assays of transcription is also found in the case of the ovalbumin gene. Thus the presence of ovalbumin-specific RNA only in the nuclear RNA of hormonally stimulated oviduct tissue is paralleled by the ability of nuclei prepared from stimulated oviduct cell nuclei to transcribe the ovalbumin gene at high levels in run-on assays. In contrast, nuclei from other tissues or unstimulated oviduct cells failed to transcribe this gene, paralleling the observed absence of ovalbumin RNA in the nuclei and cytoplasm of these tissues (Swaneck et al. 1979). Hence the observed increase in nuclear and cytoplasmic RNA for ovalbumin in response to oestrogen is indeed caused by increased transcription of the ovalbumin gene. Interestingly, in this study the incorporation of label into ovalbumin RNA in the run-on assay was observed to peak after 15 minutes and did not decrease at longer labelling times of up to 1 hour. This suggested that, unlike intact cells, isolated nuclei do not degrade or process the RNA that they synthesize, and hence allowed the use of longer labelling times, further increasing the sensitivity of this technique.

This increased sensitivity has allowed the use of nuclear run-on assays to demonstrate that transcriptional control of many genes encoding specific proteins is responsible for the previously observed differences in the levels of these proteins in different situations. Cases where such transcriptional control has been demonstrated in this manner are far to numerous to mention individually but involve a range of different tissues and organisms, such as the synthesis of α-foetoprotein in foetal but not adult mammalian liver, production of insulin in the mammalian pancreas, the expression of the *Drosophila melanogaster* yolk protein genes only in ovarian follicle cells, the expression of aggregation stage-specific genes in the slime mould *Dictyostelium discoideum,* and the expression of soya bean seed proteins such as glycinin in embryonic but not adult tissues.

These studies on individual genes for particular proteins have been supplemented by the more general studies of Darnell and colleagues (Derman *et al.* 1981, Powell *et al.* 1984). In these experiments the authors studied 12 different genes whose corresponding mRNAs were present in mouse liver cytoplasm but were absent in brain cytoplasm. These included both genes encoding previously isolated liver-specific proteins, such as albumin or transferrin, and those which had been isolated simply on the basis of the presence of their corresponding cytoplasmic RNA in the liver and not in other tissues and for which the protein product had not been identified. Measurement of the transcription rate of these genes in nuclei isolated from brain or liver showed that such transcription was detectable only in the liver nuclei (Fig. 3.8), indicating that the difference in cytoplasmic mRNA level was

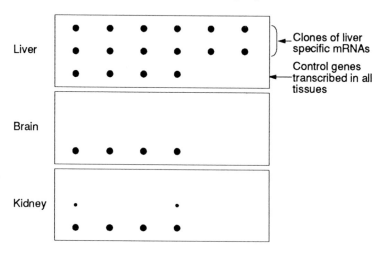

Figure 3.8 Nuclear run-on assay to measure the transcription rates of genes encoding liver-specific mRNAs and of control genes expressed in all tissues. Note that the liver specific genes are not transcribed at all in brain, while in the kidney the only two of these genes to be transcribed are those that are known to produce a low level of mRNA in the kidney.

produced by a corresponding difference in gene transcription. In these experiments the rate of transcription of the 12 genes was also measured in nuclei prepared from kidney tissue. In this tissue mRNAs corresponding to two of the genes were present at a considerably lower level than that observed in the liver, while RNA corresponding to the other 10 genes was undetectable. As with the brain nuclei, the level of transcription detectable in the kidney nuclei exactly paralleled the level of RNA present. Thus, only the two genes producing cytoplasmic mRNA in the kidney were detectably transcribed in this tissue and the level of transcription of these genes was much lower than that seen in the liver nuclei.

These studies on a relatively large number of different liver RNAs of different abundances, when taken together with the studies of many genes encoding specific proteins, indicate that for genes expressed in one or a few cell types, increased levels of RNA and protein in a particular cell type are brought about primarily by increases in gene transcription and that, at least in mammals, post-transcriptional control mechanisms such as that postulated by Davidson & Britten (1979) are not the primary means of regulating gene expression, although they may be more important in other animals, such as the sea urchin (see Ch. 4, Section 4.1).

3.2.4   *Evidence from polytene chromosomes*

Although pulse-labelling and nuclear run-on studies of many genes have conclusively established the existence of transcriptional regulation, it is necessary to discuss another means of demonstrating such regulation, in which increased transcription can be directly visualized. As described in Chapter 2 (Section 2.3.2), the chromosomal DNA in the salivary glands of *Drosophila* is amplified many times, resulting in a giant polytene chromosome. Such chromosomes exhibit along their length areas known as puffs in which the DNA has decondensed into a more open state, resulting in the expansion of the chromosome (Fig. 3.9). If cells are allowed to incorporate labelled ribonucleotides into RNA and the resulting RNA is then hybridized back to the polytene chromosomes, it localizes primarily to the positions of the puffs. Hence these puffs represent sites of intense transcriptional activity which, because of the large size of the polytene chromosomes, can be directly visualized. Most interestingly, many procedures which in *Drosophila* result in the production of new proteins, such as exposure to elevated temperature (heat shock) or treatment with the steroid hormone ecdysone, also result in the production of new puffs at specific sites on the polytene chromosomes, each treatment producing a different specific pattern of puffs. This suggests that these sites contain the

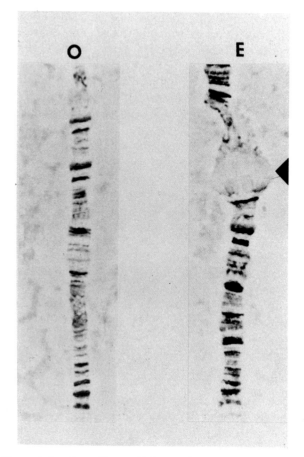

Figure 3.9 Transcriptionally active puff (arrowed) in a polytene chromosome of *Drosophila melanogaster*. The puff appears in response to treatment with the steroid hormone ecdysone (E) and is not present prior to hormone treatment (O).

genes encoding the proteins whose synthesis is increased by the treatment, and that this increased synthesis is mediated via increased transcription of these genes, which can be visualized in the puffs. In the case of ecdysone treatment this has been confirmed directly by showing that the radioactive RNA synthesized immediately after ecdysone treatment hybridizes strongly to the ecdysone-induced puffs but not to a puff which regresses upon hormone treatment. In contrast, RNA prepared from cells prior to ecdysone treatment hybridizes only to the hormone-repressed puff and not to the hormone-induced puffs (Fig. 3.10; Bonner & Pardue 1977). Similarly, RNA labelled after heat shock hybridizes intensely to puff 87C, which appears following exposure to elevated temperature and is now known to contain the

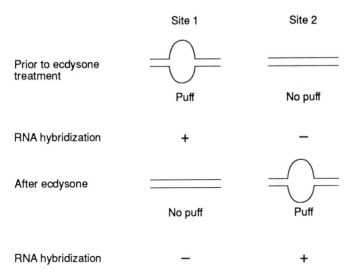

Figure 3.10 The newly synthesized RNA made following ecdysone stimulation can be labelled with [³H]uridine and shown to hybridize to the puffs that form following ecdysone treatment. Conversely, a puff that regresses after hormone treatment binds only the labelled RNA synthesized before addition of the hormone.

gene encoding the 70 KDa heat-shock protein (hsp70) which is the major protein made in *Drosophila* following heat shock (Spradling *et al.* 1975).

Thus the large size of polytene chromosomes allows a direct visualization of the transcriptional process and indicates that, as in other situations, gene activity in the salivary gland is regulated at the level of transcription.

3.3 REGULATION AT TRANSCRIPTIONAL ELONGATION

In the vast majority of cases where increased transcription of a particular gene has been demonstrated, it is likely that such increased transcription is mediated by an increased rate of initiation of transcription by RNA polymerase. Hence in a tissue in which a gene is being transcribed actively, a large number of polymerase molecules will be moving along the gene at any particular time, resulting in the production of a large number of transcripts. Such a series of nascent transcripts being produced from a single transcription unit can be visualized in the lampbrush chromosomes of amphibian oocytes, the nascent transcripts associated with each RNA polymerase molecule increasing in length the further the polymerase has proceeded along

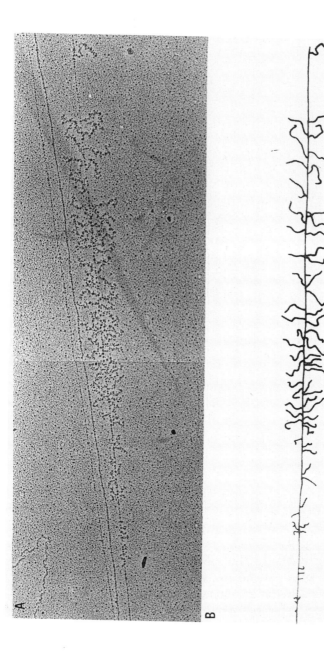

Figure 3.11 Electron micrograph (A) and summary diagram (B) of a lampbrush chromosome in amphibian oocytes, showing the characteristic nested appearance produced by the nascent mRNA chains attached to transcribing RNA polymerase molecules. The bar indicates 1 μm.

the gene, resulting in the characteristic nested appearance (Fig. 3.11).

By contrast, in tissues where a gene is transcribed at very low levels, initiation of transcription will be a rare event and only one or a very few polymerase molecules will be transcribing a gene at any particular time. Similarly, the absence of transcription of a particular gene in some tissues will result from a failure of RNA polymerase to initiate transcription in that tissue (Fig. 3.12). A large number of sequences upstream of the point at which initiation occurs and which are involved in its regulation have now been described, and these will be discussed in Chapter 6.

Interestingly, however, another means of transcriptional regulation appears to be responsible for the tenfold decline in mRNA levels encoding the cellular oncogene c-*myc* (for discussion of cellular oncogenes see Ch. 8) which occurs when the human pro-myeloid cell line HL-60 is induced to differentiate into a granulocyte type cell. If nuclear run-on assays are carried out using nuclei from undifferentiated or differentiated HL-60 cells, the results obtained vary depending on the region of the c-*myc* gene whose transcription is being measured (Bentley & Groudine 1986). Thus the c-*myc* gene consists of three exons, which appear in the messenger RNA and which are separated by intervening sequences that are removed from the primary transcript by RNA splicing. If the labelled products of the nuclear run-on procedure are hybridized to the DNA of the second exon, the levels of

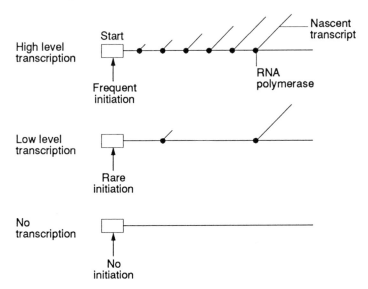

Figure 3.12 Regulation of transcriptional initiation results in differences in the number of RNA polymerase molecules transcribing a gene and therefore in the number of transcripts produced.

transcription observed are approximately tenfold higher in nuclei derived from undifferentiated cells than in nuclei from differentiated cells. Hence the observed differences in c-*myc* RNA levels in these cell types are indeed produced by differences in transcription rates. However, if the same labelled products are hybridized to DNA from the first exon of the c-*myc* gene, virtually no difference in the level of transcription of this region in the differentiated compared to the undifferentiated cells is observed. Comparison of the rates of transcription of the first and second exons in undifferentiated and differentiated cells indicates that regulation takes place at the level of transcriptional elongation, rather than initiation. Thus, although similar numbers of polymerase molecules initiate transcription of the c-*myc* gene in both cell types, the majority terminate in differentiated cells near the end of exon 1, do not transcribe the remainder of the gene, and hence do not produce a functional RNA. In contrast, in undifferentiated cells most polymerase molecules that initiate transcription, transcribe the whole gene and produce a functional RNA. Hence the fall in c-*myc* RNA in differentiated cells is regulated by means of a block to elongation of nascent transcripts (Fig. 3.13). Recently a 180 bp sequence from the 3' end of the first exon of c-*myc* has been shown to mediate this effect and to block transcriptional elongation if placed within the transcribed region of another gene (Wright & Bishop 1989). Interestingly, the rapid inhibition of transcriptional elongation following differentiation is supplemented, several days after differentiation, by an inhibition of c-*myc* transcription at the level of initiation.

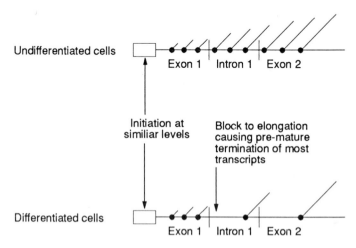

Figure 3.13 Regulation of transcriptional elongation in the c-*myc* gene. A block to elongation at the end of exon 1 results in most transcripts terminating at this point in differentiated HL-60 cells.

Similar effects on transcriptional elongation have been seen in the cellular oncogene c-*myb* and that encoding the receptor for epidermal growth factor, indicating that this mechanism is not confined to a single gene and may be quite widespread.

3.4   CONCLUSIONS

The experiments described in this chapter provide conclusive evidence that transcriptional control is the primary means used to regulate gene expression in eukaryotic organisms. The mechanism by which transcriptional control is achieved will be discussed in Chapters 5, 6 and 7. Some cases of post-transcriptional control have been described, however, and these will be discussed in the next chapter.

REFERENCES

Bentley, D. L. & M. Groudine 1986. A block to elongation is largely responsible for decreased transcription of c-*myc* in differentiated HL 60 cells. *Nature* **321**, 702–6.

Bonner, J. J. & M. L. Pardue 1977. Ecdysone-stimulated RNA synthesis in salivary glands of *Drosophila melanogaster* assay by *in situ* hybridization. *Cell* **12**, 219–25.

Darnell, J. E. 1982. Variety in the level of gene control in eukaryotic cells. *Nature* **297**, 365–71.

Davidson, E. H. & R. J. Britten 1979. Regulation of gene expression: possible role of repetitive sequences. *Science* **204**, 1052–9.

Derman, E., K. Krauter, L. Walling, C. Weinberger, M. Ray & J. E. Darnell, Jr 1981. Transcriptional control in the production of liver specific mRNAs. *Cell* **23**, 731–9.

Gilmour, R. S., P. R. Harrison, J. D. Windass, N. A. Affara & J. Paul 1974. Globin messenger RNA synthesis and processing during haemoglobin induction in Friend cells. 1. Evidence for transcriptional control in clone M2. *Cell Differentiation* **3**, 9–22.

Goldberg, R. B., G. Hosheck, G. S. Ditta & R. W. Breidenbach 1981. Developmental regulation of cloned superabundant mRNAs in soybean. *Developmental Biology* **83**, 218–31.

Groudine, M., M. Peretz & H. Weintraub 1981. Transcriptional regulation of hemoglobin switching in chicken embryos. *Molecular and Cellular Biology* **1**, 281–8.

Groudine, M., H. Hoitzer, K. Scherner & A. Therwath 1974. Lineage dependent transcription of globin genes. *Cell* **3**, 243–7.

Jacob, F. & J. Monod 1961. Genetic regulatory mechanisms in the synthesis of proteins. *Journal of Molecular Biology* **3**, 318–56.

Jeffreys, A. J. & R. A. Flavell 1977. The rabbit β-globin gene contains a large insert in the coding sequence. *Cell* **12**, 1097–108.

Kamaly, J. C. & R. B. Goldberg 1980. Regulation of structural gene expression in tobacco. *Cell* **19**, 935–46.

Latchman, D. S., H. Brzeski, R. H. Lovell-Badge & M. J. Evans 1984. Expression of the alpha-foetoprotein gene in pluripotent and committed cells. *Biochemica et Biophysica Acta* **783**, 130–6.

Lowenhaupt, K., C. Trent & J. B. Lingrel 1978. Mechanisms for accumulation of globin mRNA during dimethyl sulfoxide induction of mouse erythro-leukaemia cells: synthesis of precursors and mature mRNA. *Developmental Biology* **63**, 441–54.

Nevins, J. R. 1983. The pathway of eukaryotic mRNA transcription. *Annual Review of Biochemistry* **52**, 441–6.

Powell, D. J., J. M. Freidman, A. J. Oulethe, K. S. Krauter & J. E. Darnell, Jr 1984. Transcriptional and post-transcriptional control of specific messenger RNAs in adult and embryonic liver. *Journal of Molecular Biology* **179**, 21–35.

Roop, D. R., J. L. Nordstrom, S.-Y. Tsai, M.-J. Tsai & B. W. O'Malley 1978. Transcription of structural and intervening sequences in the ovalbumin gene and identification of potential ovalbumin mRNA precursors. *Cell* **15**, 671–85.

Sharp, P. A. 1987. Splicing of messenger RNA precursors. *Science* **235**, 766–71.

Spradling, A., S. Penman & M. L. Pardue 1975. Analysis of *Drosophila* mRNA by *in situ* hybridization: sequences transcribed in normal and heat shocked cultured cells. *Cell* **4**, 395–404.

Swaneck, G. E., J. L. Nordstrom, F. Kreuzaler, M.-J. Tsai & B. W. O'Malley 1979. Effect of oestrogen on gene expression in chicken oviduct, evidence for transcriptional control of the ovalbumin gene. *Proceedings of the National Academy of Sciences of the USA* **76**, 1049–53.

Tilghman, S. M., P. J. Curtis, D. C. Tiemeier, P. Leder & C. Weissmann 1978. The intervening sequence of a mouse beta-globin gene is transcribed within the 15S beta-globin mRNA precursor. *Proceedings of the National Academy of Sciences of the USA* **75**, 1309–13.

Wold, B. J., W. H. Klein, B. R. Hough-Evans, R. J. Britten & E. H. Davidson 1978. Sea urchin embryo mRNA sequences expressed in the nuclear RNA of adult tissues. *Cell* **14**, 941–50.

Wright, S. & J. M. Bishop 1989. DNA sequences that mediate attenuation of transcription from the mouse proto-oncogene c-*myc*. *Proceedings of the National Academy of Sciences of the USA* **86**, 505–9.

CHAPTER FOUR

Post-transcriptional regulation

4.1 REGULATION AFTER TRANSCRIPTION?

Although the evidence discussed in the preceding chapter indicates that, in mammals at least, the primary control of gene expression lies at the level of transcription, a number of cases exist where changes in the rate of synthesis of a particular protein occur without a change in the transcription rate of the corresponding gene. Indeed, in some lower organisms post-transcriptional regulation may constitute the predominant form of regulation of gene expression.

In the sea urchin, for example, the nuclear RNA contains many more different RNA species than are found in the cytoplasmic messenger RNA. Hence a large proportion of the genes transcribed give rise to RNA products that are not transported to the cytoplasm and do not function as a messenger RNA. Interestingly, however, this process is regulated differently in different tissues; an RNA species which is confined to the nucleus in one tissue being transported to the cytoplasm and functioning as a messenger RNA in another tissue. Thus, up to 80 per cent of the cytoplasmic mRNAs found in the embryonic blastula are absent from the cytoplasmic RNA of adult tissues, such as the intestine, but are found in the nuclear RNA of such tissues (Wold *et al.* 1978).

Although post-transcriptional regulation in mammals does not appear to be as generalized as in the sea urchin, some cases exist where changes in cytoplasmic mRNA levels occur without alterations in the rate of gene transcription. Such post-transcriptional regulation may be more important in controlling variations in the level of mRNA species expressed in all tissues than in the regulation of mRNA species that are expressed in only one or a few tissues. For example, in the experiments of Darnell and colleagues (Powell *et al.* 1984), which demonstrated the importance of transcriptional control in the regulation of liver-specific mRNAs (see Ch. 3, Section 3.2.3), tissue-specific differences in the levels of the mRNAs encoding actin and tubulin (which are expressed in all cell types) were observed in the absence of differences in transcription rates. Clearly, these and other cases where mRNA levels alter in the absence of changes in transcription rates indicate the existence of post-transcriptional control processes and require an understanding of their mechanisms.

In principle, such post-transcriptional regulation could operate at any of the many stages between gene transcription and the translation of the corresponding mRNA in the cytoplasm. Indeed, the available evidence indicates that in different cases regulation can occur at any one of these levels. Each of these will now be discussed in turn.

4.2   REGULATION OF RNA SPLICING

4.2.1   RNA splicing

The finding that the protein-coding regions of eukaryotic genes are split by intervening sequences (introns) which must be removed from the initial transcript by RNA splicing of the protein-coding exons (Fig. 4.1; for a recent review of this topic see Sharp 1987) led to much speculation that this process might provide a major site of gene regulation. Thus, in theory, an RNA species transcribed in several tissues might be correctly spliced to yield functional RNA in one tissue and remain unspliced in another tissue. An unspliced RNA would either be degraded within the nucleus or, if transported to the cytoplasm, would be unable to produce a functional protein due to the interruption of the protein-coding regions (Fig. 4.2). Several cases of such processing versus discard decisions have now been described in *Drosophila* (reviewed by Bingham *et al.* 1988). One decision of this type

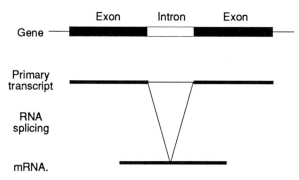

Figure 4.1 Removal of an intervening sequence from the primary RNA transcript by RNA splicing.

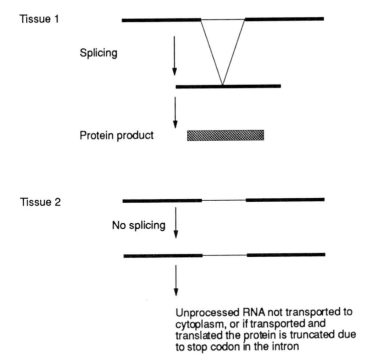

Figure 4.2 The absence of RNA splicing of a transcript in a particular tissue results in a lack of production of the corresponding protein.

is involved in regulating the movement of the transposable P element in *Drosophila*. This DNA element encodes a protein which allows it to move from position to position within the genome, but this process occurs only in the germ cells and not in somatic tissue. Detailed analysis of the gene encoding this protein (Fig. 4.3) has shown that to produce a functional mRNA three intervening sequences must be removed from the initial transcript. The third of these intervening sequences is removed only in germ cells so that somatic cells accumulate a non-functional, partially spliced transcript in which only the first two intervening sequences have been removed.

4.2.2 *Alternative RNA splicing*

Although no clear case of a processing versus discard decision has as yet been defined in mammals, numerous cases of alternative RNA splicing have been described, both in mammals and other organisms. In this process (for a review see Leff *et al.* 1986) a single gene is transcribed in several different tissues, the transcripts from this gene being processed differentially to yield different functional messenger

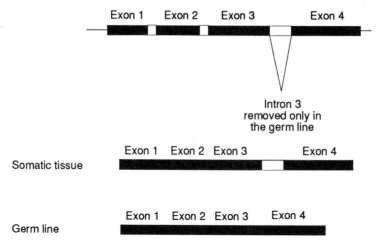

Figure 4.3 Removal of intron 3 of the P-element transposase transcript occurs only in germ line tissue.

RNAs in the different tissues (Fig. 4.4). In many cases these RNAs are translated to yield different protein products. It is noteworthy that this mechanism of gene regulation involves not only regulation of processing but also regulation of transcription, in that the alternatively processed RNAs are transcribed in only a restricted range of cell types and not in many other cells.

Cases of alternative RNA processing occur in the genes involved in a wide variety of different cellular processes, ranging from genes which regulate embryonic development or sex determination in *Drosophila* to those involved in muscular contraction or neuronal function in mammals. A representative selection of such cases is given in Table 4.1.

For convenience, cases of alternative RNA processing can be divided into three groups:

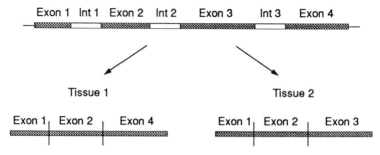

Figure 4.4 Alternative splicing of the same primary transcript in two different ways results in two different mRNA molecules.

(a) Situations where the 5' end of the differentially processed transcripts is different.

(b) Situations where the 3' end of the differentially processed transcripts is different.

(c) Situations where both the 5' and 3' ends of the differentially processed transcripts are identical.

SITUATIONS WHERE THE 5' END OF THE TRANSCRIPTS IS DIFFERENT

In these cases, two alternative primary transcripts are produced by transcription from different promoter elements and these are then processed differentially. In several situations, such as the mouse α-amylase gene (Fig. 4.5), differential splicing is controlled simply by the presence or absence of a particular exon in the primary transcript. Thus in the salivary gland, where transcription takes place from an upstream promoter, the exon adjacent to this promoter is included in the processed RNA and a downstream exon is omitted. In the liver, where the transcripts are initiated 2.8 kb downstream and do not contain the upstream exon, the processed RNA includes the downstream exon (Young *et al.* 1981). However, other cases of this type, for example that of the myosin light chain gene (Fig. 4.6), are more complex with each of the alternative primary transcripts containing both the alternatively spliced exons. In such cases it is assumed that the different primary transcripts fold into different secondary structures which favour the different splicing events.

Whether this is the case or not, it is clear that cases of alternative

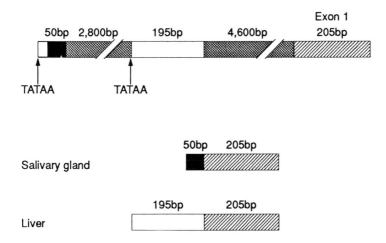

Figure 4.5 Alternative splicing at the 5' end of α-amylase transcripts in the liver and salivary gland. The two alternative start sites for transcription are indicated (TATAA), together with the 5' region of the mRNAs produced in each tissue.

71

Table 4.1

Cases of alternative splicing which are regulated developmentally or tissue specifically

Protein	Species	Nature of transcripts which undergo alternative splicing	Cell types carrying out alternative splicing
a) <u>Immune system</u>			
Immunoglobulin heavy chain IgD, IgE, IgG, IgM	Mouse	3' end differs	B cells
Lyt - 2	Mouse	Same transcript	T cells
b) <u>Enzymes</u>			
Alcohol dehydrogenase	*Drosophila*	5' end differs	Larva and adult
Aldolase A	Rat	5' end differs	Muscle and liver
α -Amylase	Mouse	5' end differs	Liver and salivary gland
(2'5') oligo A synthetase	Human	3' end differs	B cells and monocytes
c) <u>Muscle</u>			
Myosin light chain	Rat/Mouse/ Human /Chicken	5' end differs	Cardiac and smooth muscle
Myosin heavy chain	*Drosophila*	3' end differs	Larval and adult muscle
Tropomyosin	Mouse/Rat/ Human/ *Drosophila*	Same transcript	Different muscle cell types
Troponin T	Rat/Quail/ Chicken	Same transcript	Different muscle cell types

d) <u>Nerve cells</u>

Calcitonin/CGRP	Rat/Human	3' end differs	Thyroid C cells or neural tissue
Myelin Basic Protein	Mouse	Same transcript	Different glial cells
Neural cell adhesion molecule	Chicken	Same transcript	Neural development
Preprotachykinin	Bovine	Same transcript	Different neurons

e) <u>Others</u>

Fibronectin	Rat/Human	Same transcript	Fibroblasts and hepatocytes
Early retinoic acid induced gene 1	Mouse	Same transcript	Stages of embryonic cell differentiation
Thyroid hormone receptor	Rat	Same transcript	Different tissues

splicing arising from differences in the site of transcriptional initiation represent further examples of transcriptional regulation in which the variation in RNA splicing is secondary to the selection of the different promoters in the different tissues. This is not so for the remaining two categories of alternative processing event.

SITUATIONS WHERE THE 3' END OF THE TRANSCRIPTS IS DIFFERENT

After the primary transcript has been produced, it is rapidly cleaved at a point downstream of the protein-coding information and a run of adenosine residues (the poly A tail) is added post-transcriptionally. In many genes the process of cleavage and polyadenylation occurs at a different position within the primary transcript in different tissues, and the different transcripts are then differentially spliced.

The best-defined example of this process occurs in the genes encoding the immunoglobulin heavy chain of the antibody molecule, and plays an important role in the regulation of the antibody response to infection. Thus, early in the immune response, the antibody-producing B-cell synthesizes membrane-bound immunoglobulin molecules whose interaction with antigen triggers proliferation of the B-cell and results in the production of more antibody-synthesizing cells. The

73

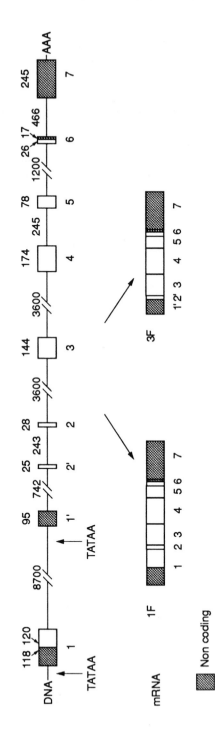

Figure 4.6 Alternative splicing of the myosin light chain transcripts in different muscle cell types produces two mRNAs (1F and 3F) differing at their 5' ends. The two alternative start sites of transcription used to produce each of the RNAs are indicated (TATAA), together with the intron–exon structure of the gene.

immunoglobulin produced by these cells is secreted, however, and can interact with antigen in tissue fluids, triggering the activation of other cells in the immune system. The production of membrane-bound and secreted immunoglobin molecules is controlled by the alternative splicing of different RNA molecules differing in their 3' ends (Fig. 4.7). The longer of these two molecules contains two exons encoding the portion of the protein that anchors it in the membrane. When this molecule is spliced, both these two exons are included, but a region encoding the last 20 amino acids of the secreted form is omitted. In the shorter RNA, the two transmembrane domain-encoding exons are absent and the region specific to the secreted form is included in the final messenger RNA.

If the polyadenylation site used in the production of the shorter immunoglobulin RNA is artifically removed, preventing its use (Danner & Leder 1985), the expected decrease in the production of secreted immunoglobulin is paralleled by a corresponding increase in the synthesis of the membrane-bound form of the protein (Fig. 4.8A). This indicates that the choice of splicing pattern is controlled by

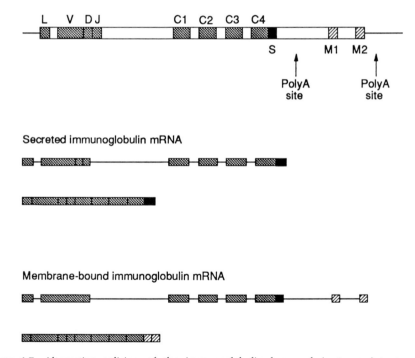

Figure 4.7 Alternative splicing of the immunoglobulin heavy chain transcript at different stages of B-cell development. The two unspliced RNAs produced by use of the two alternative polyadenylation sites in the gene are shown, together with the spliced mRNAs produced from them.

A) Immunoglobulin heavy chain

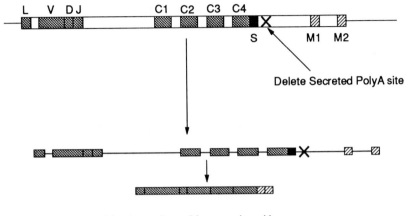

B) Calcitonin / CGRP

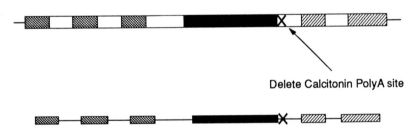

Figure 4.8 Effect of deleting the more upstream of the two polyadenylation sites in the immunoglobulin heavy chain (A) and calcitonin/CGRP genes (B) on the production of the alternatively spliced RNAs derived from each of these genes.

which polyadenylation site is used; removal of the upstream site resulting in increased use of the downstream site and increased production of the messenger RNA encoding the membrane-bound form.

This finding indicates that in at least some cases of this type the primary regulatory event to be understood is that determining the site of cleavage and polyadenylation, and that, as with cases of differential

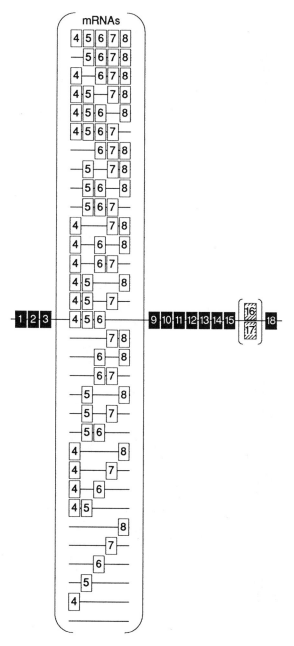

Figure 4.10 Alternative splicing of the four combinatorial exons (4–8) and the two mutually exclusive exons (16 and 17) can result in up to 64 distinct mRNAs from the rat troponin T gene.

CGRP mRNA in tissues normally producing calcitonin suggests that particular splicing factor(s) expressed only in CGRP-producing tissues are required for the production of the CGRP mRNA.

The possible existence of such factors has been further investigated by removing the promoter of the calcitonin/CGRP gene (which functions only in a restricted range of cells) and replacing it with the metallothionein promoter, which functions in all cell types. This construct was then introduced into a fertilized mouse egg *in vitro*. Two-cell embryos developing from these eggs were re-implanted into foster mothers and allowed to develop. This resulted in the eventual production of a transgenic animal in which every cell was expressing the calcitonin/CGRP gene. In such animals the vast majority of tissues produced only the spliced calcitonin mRNA, while only heart and brain (the natural site of CGRP production) produced the spliced CGRP mRNA.

These experiments strongly indicate that although all tissues have the capacity to produce calcitonin mRNA, an extra factor found only in some tissues is required to produce the CGRP mRNA. This factor appears to act by inhibiting the splicing event required to produce calcitonin mRNA, allowing the production of the CGRP mRNA (Emeson *et al*. 1989).

The evidence in favour of the existence of tissue-specific splicing factors has led to a search for tissue-specific variations in the constituents of the multicomponent spliceosome which catalyses RNA splicing. This structure contains both proteins and small RNA molecules ranging from 56 to 217 bases in size and known as the U RNAs (reviewed by Maniatis & Reed 1987). Although the majority of proteins and RNAs of the spliceosome are apparently expressed in all tissues, variants of the U1 RNA which are expressed only in certain tissues or stages of development have been reported in both mammals and Amphibia (see, for example, Lund *et al*. 1985). However, the expression of a specific variant of U1 in certain tissues has not thus far been correlated with a particular pattern of alternative RNA splicing in these tissues.

However, such a correlation has been reported in the case of a protein component of the spliceosome (McAllister *et al*. 1988, Sharpe *et al*. 1989). A novel splicing protein has been identified whose expression amongst mouse tissues is confined to brain and heart; precisely the tissues which, in the experiments described above, were able to produce correctly spliced CGRP messenger RNA. The protein is also detectable in several cell lines that can carry out the CGRP-specific splicing event, but not in a variety of others that can produce only calcitonin mRNA. These findings strongly implicate this protein as a tissue-specific splicing factor required for CGRP production. Its detection in the heart, which does not normally transcribe the calcitonin/CGRP gene, suggests that it may also be involved in other, as yet uncharacterized, cases of alternative splicing involving this tissue. It seems probable that this protein is one of several such tissue-specific splicing factors whose regulated activity is involved in controlling the many cases of alternative splicing which have been described.

4.2.4   Generality of alternative RNA splicing

The cases discussed above indicate the use of alternative splicing in a wide variety of biological processes. In mammals such splicing has been shown to regulate the immune system's production of antibodies, the production of neuropeptides such as CGRP and the tachykinins, substance P and substance K, as well as the synthesis of the different forms of at least four of the eight major sarcomere muscle proteins. Similarly, in *Drosophila* much of the posterior body plan is determined by developmentally regulated differential splicing of the ultrabithorax gene, while sex determination is also controlled by differential splicing of a hierarchy of genes in males and females (reviewed by Baker 1989).

The widespread use of alternative splicing in mammals does not refute the initial conclusion that gene regulation occurs primarily at the level of transcription, however. Thus, alternative splicing represents a response to a requirement for the production of related but different forms of a gene product in different tissues. It therefore supplements the regulation of transcription of the gene responsible for producing the different forms. Thus the immunoglobulin heavy chain gene, which produces both membrane-bound and secreted forms of the protein at different stages of B-cell development, is transcribed only in B-cells and not in other cell types, while the transcription of the troponin T gene, which produces multiple different isoforms in different muscle cell types, is confined to differentiated muscle cells. None the less, alternative splicing represents a significant supplement to the regulation of transcription, and the flexibility it confers suggests that many more cases will be identified in the future.

4.2.5   RNA editing

The finding that two different protein products can be produced from the same RNA by alternative splicing has recently been supplemented by the observation (Powell *et al.* 1987) that a similar result can be achieved by a post-transcriptional sequence change in the messenger RNA. Thus apolipoprotein B, which plays an important role in lipid transport, is known to exist in two closely related forms. A large

Apolipoprotein

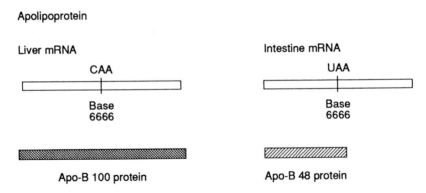

Figure 4.11   RNA editing of the apolipoprotein B transcript in the intestine produces an mRNA encoding the truncated protein apo-B48.

protein of 512 kDa known as apo-B100 is synthesized by the liver, while a smaller protein apo-B48 is made by the intestine. The smaller protein is identical to the amino-terminal portion of the larger protein. Analysis of the mRNA encoding these proteins revealed a 14.5 kb RNA in both tissues. These two RNAs were identical with the exception of a single base at position 6666, which is a cytosine in the liver transcript and a uridine in the intestinal transcript (Fig. 4.11). This change has the effect of replacing a CAA codon, which directs the insertion of a glutamine residue, with a UAA stop codon, which causes termination of translation of the intestine RNA and hence results in the smaller protein being made.

Only one gene encoding these proteins is present in the genome, and it is not alternatively spliced. In both intestinal and liver DNA this gene has a cytosine residue at position 6666. Hence the uridine in the intestinal transcript must be introduced by some form of post-transcriptional RNA editing mechanism. No information is available as to how this process occurs or even whether it occurs in the nucleus or cytoplasm, but it seems unlikely that the use of this novel mechanism in gene regulation will be confined to this single case.

4.3   REGULATION OF RNA TRANSPORT

The process of RNA splicing takes place within the nucleus, whereas the machinery for translating the spliced RNA is found in the cytoplasm. Hence the spliced mRNA must be transported to the cytoplasm if it is to direct protein synthesis. Such transport is known to occur through pores in the nuclear membrane and is an energy-dependent, active process and therefore potentially regulatable. The first example of regulation at this level has been described recently (Malim *et al.* 1989). Thus the Rev protein of human immunodeficiency virus has been shown to accelerate transport of viral RNA from nucleus to cytoplasm. Although this form of regulation may be confined to the eukaryotic viruses, it seems probable that cases involving cellular genes will also be identified in the future. If such a mechanism does act on cellular RNAs, it would, for example, provide a convenient explanation of the existence of RNA species that are confined to the nucleus in specific tissues of the sea urchin (Section 4.1). Similarly, a mechanism which prevented the transport of unspliced RNA in genes regulated by processing versus discard decisions (see Section 4.2) would represent a means of preventing wasteful translation of RNA species containing interruptions in the protein-coding sequence.

4.4   REGULATION OF RNA STABILITY

4.4.1   Cases of regulation by alterations in RNA stability

Once the mRNA has entered the cytoplasm, the number of times that it is translated, and hence the amount of protein it produces, will be determined by its stability. The more rapidly degraded an RNA is, the less protein it will produce. Hence an effective means of gene

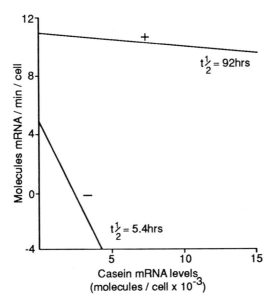

Figure 4.12   Difference in stability of the casein mRNA in the presence (+) or absence (−) of prolactin.

83

regulation could be achieved by changing the stability of an RNA species in response to some regulating signal. A number of situations where the stability of a specific RNA species is changed in this way have been described (for reviews of this topic see Brawerman 1987, Raghow 1987). Thus, for example, the mRNA for the milk protein, casein, turns over with a half-life of around 1 hour in untreated mammary gland cells. Following stimulation with the hormone

Table 4.2

Regulation of RNA stability

mRNA	Cell Type	Regulatory event	Increase or decrease in half life
Cellular oncogene c-*myc*	Friend erythroleukaemia cells	Differentiation in response to DMSO	Decrease from 35 to less than ten minutes
c-*myc*	B cells	Interferon treatment	Decreased
c-*myc*	Chinese hamster lung fibroblasts	Growth stimulation	Increased
Epidermal growth factor receptor	Epidermal Carcinoma cells	Epidermal growth factor	Increased
Casein	Mammary gland	Prolactin	Increased from one hour to forty hours
Vitellogenin	Liver	Oestrogen	Increased thirty fold
Type I pro-collagen	Skin fibroblasts	Cortisol	Decreased
Type I pro-collagen	Skin fibroblasts	Transforming growth factor beta	Increased
Histone	HeLa	Cessation of DNA synthesis	Decreased from forty minutes to eight minutes
Tubulin	CHO	Accumulation of free tubulin sub units	Decreased ten fold

FSH β   pituitary   84 Testosterone   ↑.

prolactin, the half-life increases to over 40 hours (Guyette *et al.* 1979), resulting in increased accumulation of casein mRNA and protein production in response to the hormone (Fig. 4.12). Similarly, the increased production of the DNA-associated histone proteins in the S (DNA synthesis) phase of the cell cycle is regulated in part by a fivefold increase in histone mRNA stability that occurs in this phase of the cell cycle (for reviews of the regulation of histone gene expression see Schumperli 1986, Marzluff & Pandey 1988). A representative selection of cases where mRNA stability is altered in a particular situation is given in Table 4.2.

4.4.2 Mechanisms of stability regulation

The first stage in defining the mechanism of changes in RNA stability is to identify the sequences within the RNA that are involved in mediating the observed alterations. This can be achieved by transferring parts of the gene encoding the RNA under study to another gene and observing the effect on the stability of the RNA expressed from the resulting hybrid gene. In a number of cases, short regions have been identified which can confer the pattern of stability regulation of the donor gene upon a recipient gene that is not normally regulated in this manner. In many cases such regions are located in the 3' untranslated region of the mRNA, downstream of the stop codon that terminates production of the protein. Thus the cell-cycle-dependent regulation of histone H3 mRNA stability is controlled by a 30 nucleotide sequence at the extreme 3' end of the molecule. Similarly, the destabilization of the mRNA encoding the transferrin receptor in response to the presence of iron can be abolished by deletion of a 60 nucleotide sequence within the 3' untranslated region (Mullner & Kuhn 1988). Interestingly, both of these sequences have the potential to form stem-loop structures by intra-molecular base pairing (Fig. 4.13), suggesting that changes in stability might be brought about by alterations in the folding of this region of the RNA in response to a specific signal. A similar involvement of stem-loop structures in the 3' untranslated region has also been suggested to account for changes in the stability of chloroplast mRNAs during plant development (reviewed by Gruissem 1989).

The localization of sequences involved in the regulated degradation of specific mRNA species to the 3' untranslated region is in agreement with the important role of this region in determining the differences in stability observed between different RNA species (Shaw & Kamen 1986), suggesting that differences in RNA stability, whether between different RNA species or in a single RNA in different situations, may be controlled primarily by this region. Despite this, cases where other

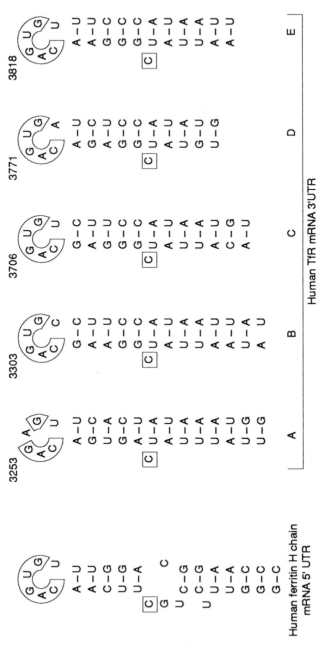

Figure 4.13 Similar stem-loop structures in the human ferritin and transferrin receptor mRNAs. Note the boxed conserved sequences in the unpaired loops and the absolute conservation of the boxed C residue, found within the stem, five bases 5' of the loop.

regions of the RNA mediate the observed alterations in stability have also been described. The most extensively studied of such cases concerns the auto-regulation of the mRNA encoding the microtubule protein, β-tubulin, in response to free tubulin monomers (Pachter *et al.* 1987, Yen *et al.* 1988). This auto-regulation prevents the wasteful synthesis of tubulin when excess free tubulin, not polymerized into microtubules, is present, and is caused by a destabilization of the tubulin mRNA. A short sequence, only 13 bases in length, from the 5′ end of the β-tubulin mRNA is responsible for this destabilization, and can confer the response on an unrelated mRNA. Most interestingly, these bases actually encode the first four amino acids of the tubulin protein, raising the possibility that the trigger for degradation of the tubulin mRNA might be the recognition of these amino acids in the tubulin protein, rather than the corresponding nucleotides in the tubulin RNA. In an elegant series of studies, Cleveland and colleagues (Yen *et al.* 1988) showed that this was indeed the case. Thus changing

a) Changing the translational reading frame by insertion of four nucleotides upstream of the normal initiation codon abolishes auto-regulation

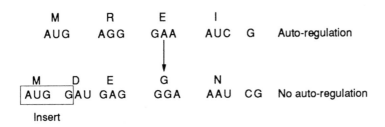

b) Changing the nucleotide sequence only abolishes auto-regulation if the encoded amino-acids are also changed

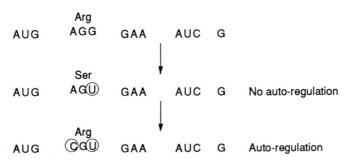

Figure 4.14 Effect of changes in the β-tubulin sequence on auto-regulation of tubulin mRNA stability.

the translational reading frame of this region, such that the identical nucleotide sequence encoded a different amino-acid sequence, abolished the auto-regulatory response (Fig. 4.14a), whereas changing the nucleotide sequence in a manner which did not alter the encoded amino acids (due to the degeneracy of the genetic code) left the response intact (Fig. 4.14b).

This finding provides an explanation for the earlier observation that the tubulin RNA must be translated (at least in part) into protein for degradation to occur. Such a requirement for translation of the RNA to be degraded is also observed, however, in cases where the target sequence for degradation lies at the 3' end of the molecule, in a region that is not translated into protein. In the case of the histone mRNA for example, introduction of stop codons resulting in premature termination of translation resulted in the abolition of the selective destabilization of the RNA following the cessation of DNA synthesis. Such an effect could also be achieved by inserting additional sequences between the normal termination codon and the previously discussed stem-loop structure which is involved in the regulation of degradation (Graves *et al.* 1987; Fig. 4.15).

These apparently contradictory observations can be combined into a model of RNA stability regulation, if it is assumed that the ribosome translating the RNA species carries with it a nuclease capable of degrading the RNA in response to a specific signal. In the case of tubulin this nuclease is triggered by a recognition event involving the first four amino acids emerging from the ribosome and then it degrades the RNA (Fig. 4.16a). In contrast, in the case of histone mRNA the trigger is supplied by recognition of the 3' RNA sequence. However, translation of the upstream part of the molecule is still required in

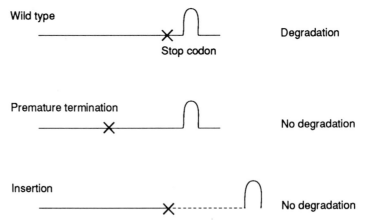

Figure 4.15 Effect of altering the relative position of the stop codon (x) and the stem-loop structure in histone mRNA on its degradation.

a) Tubulin

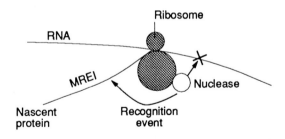

b) Histone

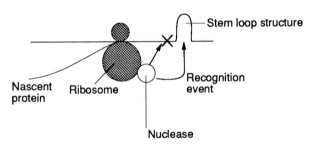

Figure 4.16 Putative mechanism for recognition of degradation signals in nascent tubulin protein (a) or histone mRNA (b) by a ribosome-associated nuclease.

order to deliver the nuclease-bearing ribosome to the appropriate part of the molecule for recognition and degradation to occur (Fig. 4.16b).

The finding that translation is also required both for the regulated degradation of other RNA species, such as the c-*myc* mRNA, as well as for the rapid turnover of short-lived mRNAs, such as that derived from another cellular oncogene, c-*fos*, suggests that this may represent a generally applicable mechanism controlling RNA degradation.

4.4.3   Role of stability changes in regulation of gene expression

A consideration of the situations where changes in the stability of a particular RNA occur (see Table 4.2) suggests that the majority have two features in common. First, changes in stability of a particular mRNA are very often accompanied by parallel alterations in the transcription rate of the corresponding gene. Thus prolactin treatment of mammary gland cells results in a two- to fourfold increase in casein gene transcription, and the increased stability of histone mRNA in the S phase of the cell cycle is accompanied by a three- to fivefold increase in transcription of the histone genes. Secondly, cases where RNA stability is regulated are very often those where a rapid and transient change in the synthesis of a particular protein is required. Thus synthesis of the histone proteins is necessary only at one particular phase of the cell cycle, when DNA is being synthesized. Following cessation of DNA synthesis a rapid shut-off in the synthesis of unnecessary histone proteins is required. Similarly, following the cessation of hormonal stimulation it would be highly wasteful to continue the synthesis of hormonally dependent proteins such as casein or vitellogenin. In the case of the cellular oncogene c-*myc*, whose RNA stability is transiently increased when cells are stimulated to grow, such continued synthesis would not only be highly wasteful but is also potentially dangerous to the cell. Thus this growth-regulatory protein is only required for a short period when cells are entering the growth phase, its continued inappropriate synthesis at other times carrying the risk of disrupting cellular growth regulatory mechanisms, possibly resulting in transition to a cancerous state (see Ch. 8).

These considerations suggest that alterations in RNA stability are used as a significant supplement to transcriptional control in cases where rapid changes in the synthesis of a particular protein are required. Thus, if transcription is shut off in response to withdrawal of a particular signal, inappropriate and metabolically expensive protein synthesis will continue for some time from pre-existing mRNA, unless that RNA is degraded rapidly. Similarly, rapid onset of the expression of a particular gene can be achieved by having a relatively high basal level of transcription with high RNA turnover in the absence of stimulation, allowing rapid onset of translation from pre-existing RNA following stimulation. Hence, as with alternative RNA processing, cases where RNA stability is regulated represent an adaptation to the requirements of a particular situation and do not affect the conclusion that regulation of gene expression occurs primarily at the level of transcription.

4.5   REGULATION OF TRANSLATION

4.5.1   *Cases of translational control*

The final stage in the expression of a gene is the translation of its messenger RNA into protein. In theory therefore, the regulation of gene expression could be achieved by producing all possible mRNA species in every cell and selecting which were translated into protein in each individual cell type. The evidence that different cell types have very different cytoplasmic RNA populations (see Ch. 1) indicates, however, that this extreme model is incorrect. None the less, the regulation of translation, such that a particular mRNA is translated into protein in one situation and not another does occur in some special cases.

The most prominent of such cases is that of fertilization. In the unfertilized egg protein synthesis is slow, but upon fertilization of the egg by a sperm a tremendous increase in the rate of protein synthesis occurs. This increase does not require the production of new mRNAs after fertilization. Rather, it is mediated by pre-existing maternal RNAs which are present in the unfertilized egg but are only translated after fertilization. Although in many species such translational control produces only quantitative changes in protein synthesis, in others it can affect the nature as well as the quantity of the proteins being made before and after fertilization. Thus in the clam *Spisula solidissima* some new proteins appear after fertilization, while others which are synthesized in large amounts before fertilization are repressed there-after. However, the RNA populations present before and after fertilization are identical (Fig. 4.17), indicating that translational control processes are operating (Rosenthal *et al.* 1980).

Similar translational control processes operating on many RNA species also occur following infection with the large DNA viruses, such as adenovirus or herpes simplex virus, and following the exposure of cells to elevated temperatures (heat shock). In both these cases, the translation of most pre-existing cellular RNA species is repressed, while the translation of either the viral mRNAs or of those encoding the heat-shock proteins (see Ch. 3) occurs at high levels.

As well as producing parallel changes in the translation of many RNA species, translational control processes may also operate on individual RNAs in a particular cell type. Thus, for example, the rate of translation of the globin RNA in reticulocytes is regulated in response to the availability of the haeme co-factor which is required for the production of haemoglobin. Similarly, the translation of the RNA encoding the iron-binding protein, ferritin, is regulated in response to the availability of iron (Casey *et al.* 1988).

4.5.2 *Mechanism of translational control*

In principle, translational regulation could operate via modifications in the cellular translational apparatus affecting the efficiency of translation of particular RNAs, or by modifications in the RNA itself which affect the way in which it is translated by the ribosome. Evidence is available

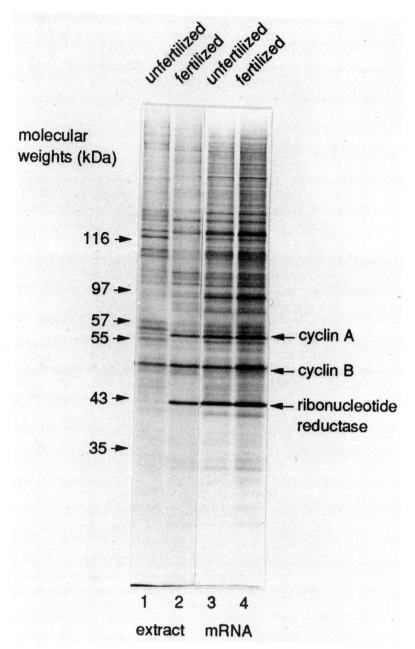

Figure 4.17 Translational control in the clam *Spisula solidissima*. Different proteins are synthesized *in vivo* before fertilization (track 1) and after fertilization (track 2). If, however, RNA is isolated either before (track 3) or after fertilization (track 4) and translated *in vitro* in a cell-free system, identical patterns of proteins are produced. Hence the difference in the proteins produced *in vivo* from identical RNA populations must be due to translational control.

indicating that both these types of mechanism are used in different cases.

Thus, in the absence of haeme a cellular protein kinase in the reticulocyte becomes active and phosphorylates the protein initiation factor eIF2, resulting in its inactivation. Since this factor is required for the initiation of protein synthesis, translation of the globin RNA ceases until haeme is available. However, the use of such a mechanism, in which total inactivation of the cellular translational apparatus is used to regulate the translation of a single RNA, is possible only in the reticulocyte, where the globin protein constitutes virtually the only translation product.

In other cell types, where a large number of different RNA species are expressed, such a mechanism can be used only where large-scale repression of many different RNA species occurs. Thus phosphorylation of eIF2 (produced by a different kinase to that activated by the absence of haeme) also occurs following infection with viruses, such as adenovirus, which inhibit the translation of most cellular mRNAs, and it is probable that a similar mechanism is also responsible for the repression of cellular protein synthesis following heat shock.

Other situations where the translation of only a single RNA species is regulated, however, are likely to operate through sequences within the specific RNA itself. Indeed, the preferential translation of the RNAs encoding the heat-shock protein, which accompanies the repression of most cellular protein synthesis following heat shock, is mediated by short sequences contained within the 5' untranslated regions of these RNAs upstream of the point at which protein synthesis is initiated (Hultmark et al. 1986). Presumably, these sequences allow the modified translational apparatus that exists after heat shock to recognize and translate these mRNAs.

Several other cases where translational regulation of particular mRNA species is mediated by sequences in their 5' untranslated region have been defined. Thus, the enhanced translation of the ferritin mRNA in response to iron is mediated by a sequence in this region which can fold into a stem-loop structure (Casey et al. 1988). The structure of this stem-loop is very similar to that found in the 3' untranslated region of the transferrin receptor mRNA, whose stability is negatively regulated by the presence of iron (see Fig. 4.13 and Table 4.3). This has led to the suggestion that such loops may represent functionally equivalent iron response elements whose opposite effects on gene expression are dependent upon their position (5' or 3') within the RNA molecule. This idea was confirmed (Casey et al. 1988) by transferring the transferrin receptor stem-loop to the 5' end of an unrelated RNA, resulting in the iron-dependent enhancement of its translation. The identical structure is therefore capable of mediating

93

Table 4.3

Regulation of the transferrin receptor and ferritin genes

	Effect of iron on protein production	Mechanism	Position of stem-loop structure
Ferritin	Increased	Increased mRNA translation	5' untranslated region
Transferrin receptor	Decreased	Decreased mRNA stability	3' untranslated region

opposite effects on RNA stability and translation, depending on its position within the RNA molecule. Such an apparent paradox can be explained if it is assumed that, in response to the presence of iron, the stem-loop structure in these mRNAs unfolds (Fig. 4.18). In the ferritin mRNA, where the element is in the 5' end, this will allow unimpeded movement of the ribosome along the mRNA and increased translation. In contrast, in the transferrin receptor mRNA, where this element is in the 3' untranslated region, such unfolding presumably renders the RNA susceptible to nuclease degradation at an increased rate. In agreement with this model, a protein whose activity increases over twentyfold in response to iron has been shown to bind to the stem-loop element in both the ferritin and transferrin receptor mRNAs (Mullner *et al.* 1989).

Not all cases of translational regulation mediated by sequences in the 5' untranslated region operate via such stem-loop structures, however. Increased expression of the yeast regulatory protein GCN4 in response to amino-acid starvation is caused by increased translation of its RNA (for a review see Fink 1986). The translational regulation of this molecule is mediated by short sequences within the 5' untranslated region of the RNA, upstream of the start point for translation of the GCN4 protein. Most interestingly, such sequences are capable of being translated to produce peptides of two or three amino acids (Fig. 4.19). Following translation of these sequences to yield such small peptides, the ribosome apparently fails to reinitiate at the translational start point for GCN4 production and hence this protein is not synthesized. Following amino-acid starvation, the production of the small peptides is suppressed and production of GCN4 correspondingly enhanced.

Interestingly, alternative initiation codons also exist in the mRNA encoding the human Ia antigen-associated invariant chain (Strubin *et al.* 1986). In this case, however, the use of each of the two possible

a) Ferritin

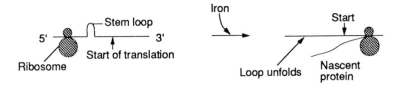

b) Transferrin receptor

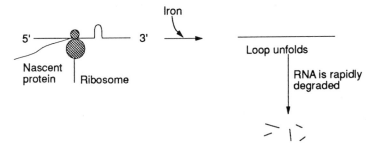

Figure 4.18 Role of iron-induced unfolding of the stem-loop structure in producing increased translation of the ferritin mRNA (panel a) and increased degradation of the transferrin receptor mRNA (panel b).

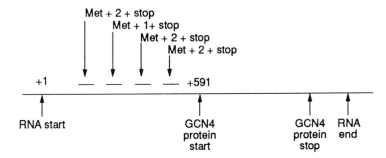

Figure 4.19 Presence of short open reading frames capable of producing small peptides in the 5′ untranslated region of the yeast GCN4 RNA. Translation of the RNA to produce these small proteins supresses translation of the GCN4 protein. The position of the methionine residue beginning each of the small peptides is indicated together with the number of additional amino acids incorporated before a stop codon is reached.

initiation codons is involved not in gene regulation but rather in the production of the two different forms of the protein found in human cells. Such alternative translation of the same mRNA parallels the use of alternative splicing to produce different protein products from the same gene.

Study of the cases of translational control where the mechanisms have been defined thus makes it clear that although sequences in the 5' untranslated region of particular mRNAs are involved in the translational regulation of their expression, the mechanism by which they do so may differ dramatically in different cases.

Although the 5' untranslated region is an obvious location for sequences involved in mediating translational control, cases where sequences in the 3' untranslated region play a role in the regulation of translation have been reported. Thus, sequences in this region are involved in modulating the efficiency of translation of the mRNA encoding β-interferon and also in mediating the increased translation of the mRNA encoding tissue plasminogen activator (t-PA), which occurs during meiotic maturation of mouse oocytes. In this latter case these sequences appear to act by affecting the addition of adenosine residues to the 3' end of the RNA, such polyadenylation being necessary for the translation of the RNA. Interestingly, sequences in the same region of the t-PA mRNA are also involved in causing the subsequent destabilization of the mRNA and its rapid degradation later in the process of oocyte maturation (Strickland et al. 1988), further emphasizing the relationship between the various forms of post-transcriptional control.

In summary, therefore, it is clear that cases of translational control can be mediated by sequences in various parts of the RNA and can involve secondary structure, use of different translation initiation codons, or the regulation of polyadenylation.

4.5.3 Significance of translational control

Many of the cases of translational control occur in situations where very rapid responses are required. Thus, following fertilization a very rapid activation of cellular growth processes is required. Similarly, following heat shock it is necessary to shut down rapidly the synthesis of most enzymes and structural proteins and begin to synthesize the protective heat-shock proteins. Such regulation can be achieved rapidly by translational control, supplemented by increased transcription of the genes encoding the heat-shock proteins. Once again, as with other cases of post-transcriptional regulation, translational control can be viewed as supplementing the regulation of transcription in order to meet the requirements of particular specialized cases. In the case of the heat-shock proteins, this combination of transcriptional and translational control is further supplemented by an increased stability of the heat-shock mRNAs following exposure to elevated temperature, providing a further means of producing a rapid and effective response (Theodorakis & Morimoto 1987).

The case of the yeast GCN4 protein provides a different aspect to translational control, however. This protein is a transcriptional regulator, which increases the transcription of several genes encoding the enzymes of amino-acid biosynthesis in response to a lack of one or more amino acids. In this case the synthesis of a transcriptional regulatory protein is regulated by translational control. When taken together with the increasing evidence (see Ch. 7) that many mammalian transcriptional regulatory molecules may pre-exist in an inactive form and be activated by protein modifications (for example phosphorylation), this suggests that it may be necessary for the cell to control the expression of some of its transcriptional regulatory molecules at levels other than that of gene transcription. Translational control may be one means by which such regulation is achieved.

4.6   CONCLUSIONS

A wide variety of cases exist in which gene expression can be regulated at levels other than transcription. In some lower organisms such post-transcriptional regulation may constitute the predominant form of gene control. In mammals, however, it appears to represent an adaptation to particular situations, including the need to generate two closely related proteins from one gene by alternative RNA splicing, or to respond rapidly to the withdrawal of hormonal stimulation or stress by the regulation of RNA stability or translation. It is likely that many more cases of post-transcriptional regulation will be described in the future, and the possibility remains that this form of regulation may be the predominant one for the proteins which themselves regulate the transcription of other genes.

REFERENCES

Baker, B. S. 1989. Sex in flies: the splice of life. *Nature* **340**, 521–4.
Bingham, P. M., T.-B. Chou, I. Mims & Z. Zachar 1988. On/off regulation of gene expression at the level of gene splicing. *Trends in Genetics* **4**, 134–8.
Brawerman, G. 1987. Determinants of messenger RNA stability. *Cell* **48**, 5–6.
Breitbart, R. E., A. Andreadis & B. Nadal-Ginard 1987. Alternative splicing: a ubiquitous mechanism for the generation of multiple protein isoforms from different genes. *Annual Review of Biochemistry* **56**, 467–95.

Casey, J. L., M. W. Hautze, D. W. Koeller, S. W. Caughman, T. A. Ronault, R. D. Klausner & J. B. Harford 1988. Iron-responsive regulatory sequences that control mRNA levels and translation. *Science* **240**, 924–8.

Danner, D. & P. Leder 1985. Role of an RNA cleavage/poly (A) addition site in the production of membrane bound and secreted IgM mRNA. *Proceedings of the National Academy of Sciences of the USA* **82**, 8658–62.

Emeson, R. B., F. Hedjran, J. M. Yeakley, J. W. Guise & M. G. Rosenfeld 1989. Alternative production of calcitonin and CGRP mRNA is regulated at the calcitonin-specific splice acceptor. *Nature* **341**, 76–80.

Fink, G. R. 1986. Translational control of transcription in eukaryotes. *Cell* **45** 155–6.

Graves, R. A., N. B. Pandey, N. Chodchoy & W. F. Marzluff 1987. Translation is required for regulation of histone mRNA degradation. *Cell* **48**, 615–26.

Gruissem, W. 1989. Chloroplast gene expression: how plants turn their plastids on. *Cell* **56**, 161–70.

Guyette, W. A., R. A. Matusik & J. M. Rosen 1979. Prolactin-mediated transcriptional and post-transcriptional control of casein gene expression. *Cell* **17**, 1013–23.

Hultmark, D., R. Klemenz & W. Gehring 1986. Translational and transcriptional control elements in the untranslated leader of the heat-shock gene hsp 22. *Cell* **44**, 429–38.

Leff, S. E., R. M. Evans, M. G. Rosenfeld 1988. Splice commitment dictates neuron-specific alternative RNA processing in calcitonin–CGRP gene expression. *Cell* **48**, 517–24.

Leff, S. E., M. G. Rosenfeld & R. M. Evans 1986. Complex transcriptional units: diversity in gene expression by alternative RNA processing. *Annual Review of Biochemistry* **55**, 1091–117.

Lund, E., B. Kahan & J. E. Dahlberg 1985. Differential control of U1 small nuclear RNA expression during mouse development. *Science* **229**, 1271–4.

McAllister, G., S. G. Amara & M. R. Lerner 1988. Tissue-specific expression and cDNA cloning of small nuclear ribonucleoprotein associated polypeptide N. *Proceedings of the National Academy of Sciences of the USA* **85**, 5296–300

Malim, M. H., J. Hauber, S.-Y. Le, J. V. Maizel & B. R. Cullen 1989. The HIV-1 rev trans-activator acts through a structured target sequence to activate nuclear export of unspliced viral mRNA. *Nature* **338**, 254–7.

Maniatis, T. & R. Reed 1987. The role of small ribonucleoprotein particles in pre-mRNA splicing. *Nature* **325**, 673–8.

Marzluff, W. F. & N. B. Pandey 1988. Multiple regulatory steps control histone mRNA concentration. *Trends in Biochemical Sciences* **13**, 49–52.

Mullner, E. W. & L. C. Kuhn 1988. A stem-loop in the 3′ untranslated region mediates iron-dependent regulation of transferrin receptor mRNA stability in the cytoplasm. *Cell* **53**, 815–25.

Mullner, E. W., B. Neupert & L. E. Kuhn 1989. A specific mRNA binding protein regulates the iron-dependent stability of cytoplasmic transferrin receptor mRNA. *Cell* **58**, 373–82.

Pachter, J. S., T. J. Yen & D. W. Cleveland 1987. Auto-regulation of tubulin expression is achieved through degradation of polysomal tubulin mRNAs. *Cell* **51**, 283–92.

Powell, D. J., J. M. Freidman, A. J. Oulette, K. S. Krauter & J. E. Darnell, Jr 1984. Transcriptional and post-transcriptional control of specific messenger RNAs in adult and embryonic liver. *Journal of Molecular Biology* **179**, 21–35.

Powell, L. M., S. C. Wallis, R. J. Pease, Y. H. Edwards, T. J. Knott & J. Scott 1987. A novel form of tissue-specific RNA processing produces apolipo-protein -B48 in intestine. *Cell* **50**, 831–40.

Raghow, R. 1987. Regulation of messenger RNA turnover in eukaryotes. *Trends in Biochemical Sciences* **12**, 358–60.

Rosenfeld, M. G., S. G. Amara & P. M. Evans 1984. Alternative RNA processing: determining neuronal phenotype. *Science* **225**, 1315–20.

Rosenthal, E. T., T. Hurt & J. V. Rudeman 1980. Selective translation of mRNA controls the pattern of protein synthesis during early development of the surf clam *Spisula solidissima*. *Cell* **20**, 487–94.

Schumperli, D. 1986. Cell-cycle regulation of histone gene expression. *Cell* **45**, 471–2.

Sharp, P. A. 1987. Splicing of messenger RNA precursors. *Science* **235**, 766–71.

Sharpe, N. G., D. G. Williams, P. M. Norton & D. S. Latchman 1989. Expression of the SmB' protein in rodent cells capable of following an alternative splicing pathway. *FEBS Letters* **243**, 132–6.

Shaw, G. & R. Kamen 1986. A conserved AU sequence from the 3' untranslated region of GM-CSF mRNA mediates selective mRNA degrada-tion. *Cell* **46**, 659–67.

Strickland, S., J. Huarte, D. Belin, A. Vassali, R. J. Rickles & J.-D. Vassali 1988. Anti-sense RNA directed against the 3' non-coding region prevents dormant mRNA activation in mouse oocytes. *Science* **241**, 680–4.

Strubin, M., E. O. Long & B. Mach 1986. Two forms of the Ia antigen-associated invariant chain result from alternative initiation at two in-phase AUGs. *Cell* **47**, 619–25.

Theodorakis, N. G. & R. I. Morimoto 1987. Post-transcriptional regulation of hsp70 in human cells: effects of heat shock, inhibition of protein synthesis and adenovirus infection on translation and mRNA stability. *Molecular and Cellular Biology* **7**, 4357–68.

Wold, B. J., W. H. Klein, B. R. Hough-Evans, R. J. Britten & E. H. Davidson 1978. Sea urchin embryo mRNA sequences expressed in the nuclear RNA of adult tissues. *Cell* **14**, 941–50.

Yen, T. J., P. S. Machlin & D. W. Cleveland 1988. Autoregulated instability of β-tubulin by recognition of the nascent amino-acids of β-tubulin. *Nature* **334**, 580–5.

Young, R. A., O. Hagenbuchle & U. Schibler 1981. A single mouse α-amylase gene specifies two different tissue specific mRNAs. *Cell* **23**, 451–8.

CHAPTER FIVE

Transcriptional control–chromatin structure

5.1 INTRODUCTION

Having established that the primary control of eukaryotic gene expression lies at the level of transcription, it is necessary to investigate the mechanisms responsible for this effect. The fact that regulation at transcription is also responsible for the control of gene expression in bacteria suggests that insights into these procedures obtained in these much simpler organisms (reviewed by Gottessman 1984) may be applicable to higher organisms. Thus it is possible to envisage that regulation of transcription in eukaryotes might occur by means of a protein, present in all tissues, which binds to the promoter region of a particular gene and prevents its expression. In one particular cell type, or in response to a particular signal such as heat shock, this protein would be inactivated either directly (Fig. 5.1a) or by binding of another factor (Fig. 5.1b), and would no longer bind to the gene. Hence transcription would occur only in the one cell type or in response to the signal. This mechanism is based on that regulating the expression of the *lac* operon containing the genes encoding proteins required for the metabolism of lactose. This operon is normally repressed by the lac repressor protein; binding of lactose to this protein, however, results in its inactivation and allows transcription of the operon (reviewed by Miller & Reznikoff 1980).

Alternatively, the fact that most eukaryotic genes are inactive in most tissues, and become active only in one tissue or in response to a particular signal suggests that it may be more economical to have a system in which the gene is constitutively inactive in most tissues, without any repressor being required. Activation of the gene would then require a particular factor binding to its promoter. The specific expression pattern of the gene would be controlled by the presence of this factor only in the expressing cell type (Fig. 5.2a) or, alternatively, by the factor's requirement for a co-factor such as a steroid hormone to

Inactive tissue

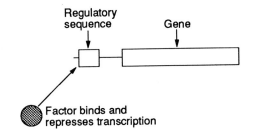

Regulatory sequence — Gene

Factor binds and represses transcription

Active tissue

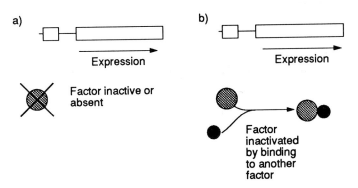

a)

Expression

Factor inactive or absent

b)

Expression

Factor inactivated by binding to another factor

Figure 5.1 Model for the activation of gene expression in a particular tissue by inactivation (a) or binding out (b) of a repressor present in all tissues.

convert it to an active form (Fig. 5.2b). This mechanism is based on the regulation of the arabinose operon in *Escherichia coli* in which binding of the substrate arabinose to the regulatory araC protein allows it to induce transcription of the genes required for the metabolism of arabinose (reviewed by Raibaud & Schwartz 1984).

Hence, based on the known mechanisms of gene regulation in bacteria, it is possible to produce models of how gene regulation might operate in eukaryotes. Indeed, a considerable amount of evidence indicates that many cases of gene regulation do use the activation type mechanisms seen in the arabinose operon. Thus, as will be discussed in Chapters 6 and 7, the effects of glucocorticoid and other steroid hormones on gene expression are mediated by the binding of the steroid to a receptor protein. This activated complex then binds to particular sequences upstream of steroid-responsive genes and acti-

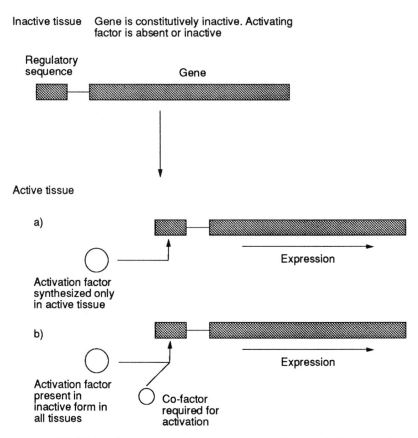

Figure 5.2 Model for the activation of gene expression in a particular tissue by an activator present only in that tissue (a) or activated by a co-factor (b) only in that tissue.

vates their transcription (reviewed by Beato 1989).

Even in the case of steroid hormones, however, such mechanisms cannot account entirely for the regulation of gene expression. In the chicken, administration of the steroid hormone oestrogen results in the transcriptional activation in oviduct tissue of the gene encoding ovalbumin, as discussed in Chapter 3. In the liver of the same organism, however, oestrogen treatment has no effect on the ovalbumin gene but instead results in the activation of a completely different gene, encoding the protein vitellogenin. Such tissue-specific differences in the response to a particular treatment are, of course, entirely absent in single-celled bacteria and cannot be explained simply on the basis of the activation of a single DNA-binding protein by the hormone.

Similarly, models of this type cannot explain the data obtained by

Becker *et al.* (1987) in their studies of the regulation of the rat tyrosine amino-transferase gene. All the protein factors binding to the regulatory regions of this gene were detectable in both the liver, where the gene is expressed, and in other tissues, where no expression is observed. Moreover, the proteins isolated from these tissues were equally active in binding to the appropriate region of the gene in deproteinized DNA. Only in the liver, however, was actual binding of these factors to the DNA of the gene within its normal chromosomal structure detectable. Once again, these data cannot be explained solely on the basis of models in which proteins stimulate or inhibit gene expression by binding to appropriate DNA sequences.

Rather, an understanding of eukaryotic gene regulation will require a knowledge of the ways in which the relatively short-term regulatory processes mediated by the binding of proteins to specific DNA sequences in this manner interact with the much longer-term regulatory processes, which establish and maintain the differences between particular tissues and also control their response to treatment with effectors such as steroids. These long-term control processes will be discussed in this chapter, and the DNA sequences and proteins which actually regulate transcription in response to particular signals will be discussed in Chapters 6 and 7.

5.2 COMMITMENT TO THE DIFFERENTIATED STATE AND ITS STABILITY

It is evident that the existence of different tissues and cell types in higher eukaryotes requires mechanisms that establish and maintain such differences, and that these mechanisms must be stable in the long term. Thus, despite some exceptions (see Ch. 2, Section 2.2.3), tissues or cells of one type do not, in general, change spontaneously into another cell type. Inspection of mammalian brain tissue does not reveal the presence of cells typical of liver or kidney, and antibody-producing B-cells do not spontaneously change into muscle cells when cultured in the laboratory. Hence, cells must be capable of maintaining their differentiated state, either within a tissue or during prolonged periods in culture.

Indeed, the long-term control processes that achieve this effect must regulate not only the ability to maintain the differentiated state more or less indefinitely but also the observed ability of cells to remember their particular cell type, even under conditions when they cannot express the characteristic features of that cell type. Thus in the experiments of Coon (1966) it was possible to regulate the behaviour of cartilage-producing cells in culture according to the medium in which they were

103

placed. In one medium the cells were capable of expressing their differentiated phenotype and produced cartilage-forming colonies synthesizing an extracellular matrix containing chondroitin sulphate. By contrast, in a medium favouring rapid division the cells did not form such colonies and, instead, divided rapidly and lost all the specific characteristics of cartilage cells, becoming indistinguishable from undifferentiated fibroblast-like cells in appearance. None the less, even after 20 generations in this rapid growth medium the cells could resume the appearance of cartilage cells and synthesis of chondroitin sulphate if returned to the appropriate medium. This process did not occur if other cell types or undifferentiated fibroblast cells were placed in an identical medium. Hence the cartilage cells were capable not only of maintaining their differentiated cell type in a particular medium supplying appropriate signals, but also of remembering it in the absence of such signals and returning to the correct differentiated state when placed in the appropriate medium.

These experiments lead to the idea that cells belong to a particular lineage and that mechanisms exist to maintain the commitment of cells to a particular lineage, even in the absence of the characteristics of the differentiated phenotype. A variety of evidence exists to suggest that cells become committed to a particular differentiated stage or lineage well before they express any features characteristic of that lineage, and that such commitment can be maintained through many cell generations. Perhaps the most dramatic example of this effect occurs in the fruit fly, *Drosophila melanogaster*. In this organism, the larva contains many discs consisting of undifferentiated cells located at intervals along the length of the body and indistinguishable from each other in appearance. Eventually, these imaginal discs (Hadorn 1968) will form the structures of the adult, the most anterior pair producing the antennae and others producing the wings, legs, etc. In order to do this, however, the discs must pass through the intermediate pupal state where they receive the appropriate signals to differentiate into the adult structures. If they are removed from the larva and placed directly in the body cavity of the adult, they will remain in an undifferentiated state, since the signals inducing differentiation will not be received (Fig. 5.3). This process can be continued for many generations. Thus the disc removed from the adult can be split in two, one half being used to propagate the cells by being placed directly in another adult and the other half being tested to see what it will produce when placed in a larva that is allowed to proceed to an adult through the pupal state. When this is done, it is found that a disc which would have given rise to an antenna can still do so when placed in a larva after many generations of passage through adult organisms in an undifferentiated state. This is the case even if the disc is placed at a position

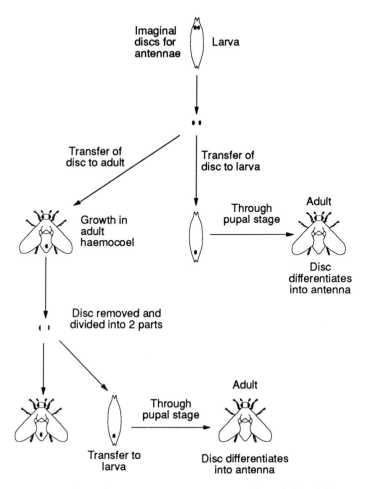

Figure 5.3 The use of larval imaginal discs to demonstrate the stability of the committed state. The disc cells maintain their commitment to produce a specific adult structure even after prolonged growth in an undifferentiated state in successive adults.

within the larva very different from that normally occupied by the imaginal disc producing the antenna. The cells of the imaginal disc within the normal larva must therefore have undergone a commitment event to the eventual production of a particular differentiated state, which they can maintain for many generations, prior to having ever expressed any features characteristic of that differentiated state.

These examples of the stability of the differentiated state and commitment to it imply the existence of long-term regulatory processes capable of maintaining such stability. However, the existence of cases where such stability breaks down, either artificially following nuclear transplantation or more naturally as in lens regeneration (see Ch. 2,

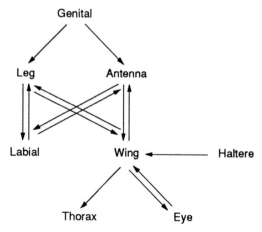

Figure 5.4 Structures that arise from specific imaginal discs of *Drosophila* following the breakdown of commitment.

Section 2.2.3), indicates that such processes, although stable, cannot be irreversible. Indeed, even in the imaginal discs of *Drosophila* the stability of the committed state can be broken down by culture for long periods in adults, the discs giving rise eventually to tissues other than the one intended originally when placed in the larva. Most interestingly, this breakdown of the committed state is not a random process but proceeds in a highly characteristic manner. Wing cells are the first abnormal cells produced by a disc that should produce an eye, with other cell types being produced only subsequently. Similarly, a disc that should produce a leg never produces genitalia, although it can produce other cell types. The various transitions that can occur in this system are summarized in Figure 5.4.

These examples and the other experiments discussed in Chapter 2 (Section 2.2) eliminate irreversible mechanisms such as DNA loss as a means of explaining the process of commitment to the differentiated state. It is thought, therefore, that the semi-stability of this process and its propagation through many cell generations is due to the establishment of a particular pattern of association of the DNA with specific proteins. The structure formed by DNA and its associated proteins is known as chromatin. An understanding of how long-term gene regulation is achieved therefore requires a knowledge of the structure of chromatin.

5.3 CHROMATIN STRUCTURE

If the DNA in a single human individual were to exist as an extended linear molecule, it would have a length of 5×10^{10} kilometres and

Table 5.1

The histones

Histone	Type	Molecular weight (kDa)	Molar ratio
H1	Lysine rich	23,000	1
H2A	Slightly lysine rich	13,960	2
H2B	Slightly lysine rich	13,774	2
H3	Arginine rich	15,342	2
H4	Arginine rich	11,282	2

would extend 100 times the distance from the Earth to the Sun (Lewin 1980). Clearly this is not the case. Rather, the DNA is compacted by folding in a complex with specific nuclear proteins into the structure known as chromatin (for a review see Igo-Kemenes *et al.* 1982). Of central importance in this process are the five types of histone proteins (Table 5.1) whose high proportion of positively charged amino acids neutralizes the net negative change on the DNA and allows folding to occur. The basic unit of this folded structure is the nucleosome (for reviews see Kornberg & Klug 1981, Morse & Simpson 1988), in which approximately 200 base pairs of DNA are associated with a histone octamer containing two molecules of each of the four core histones, H2A, H2B, H3, and H4 (Fig. 5.5). In this structure approximately 146 base pairs of DNA make almost two full turns around the histone

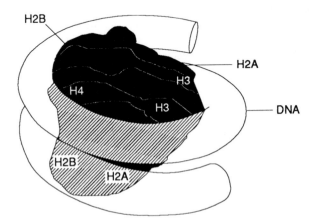

Figure 5.5 Structure of DNA and the core histones within a single nucleosome.

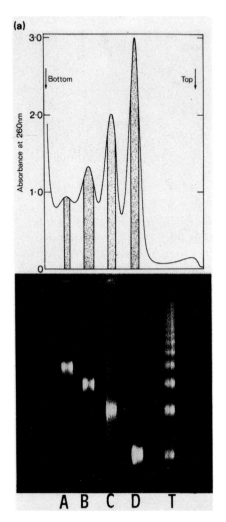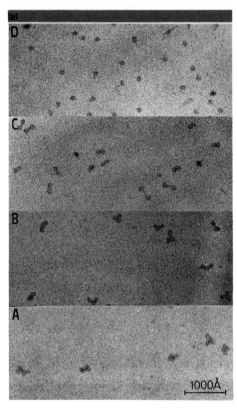

Figure 5.6 (a) Separation of mononucleosomes (D), dinucleosomes (C), trinucleosomes (B), and tetranucleosomes (A) by sucrose gradient centrifugation. The upper panel shows the peaks of absorbance produced by the individual fractions of the gradient, the lower panel shows the DNA associated with each fraction separated by gel electrophoresis. Track T shows the DNA ladder produced from a preparation containing all the individual nucleosome fractions. (b) Electron microscopic analysis of the fractions separated in panel (a). Mononucleosomes are clearly visible in fraction D, dinucleosomes in fraction C, and so on.

octamer, and the remainder serves as a linker DNA, joining one nucleosome to another.

The linker DNA is more accessible than the highly protected DNA tightly wrapped around the octamer and is therefore preferentially cleaved when chromatin is digested with small amounts of the DNA-digesting enzyme, micrococcal nuclease. Thus, if DNA is isolated following such mild digestion of chromatin, on gel electrophoresis it produces a ladder of DNA fragments of multiples of 200 base pairs, representing the results of cleavage in some but not all linker regions (Fig. 5.6a, track T). This can be correlated with the properties of nucleosomes isolated from the digested chromatin. Thus a partially digested chromatin preparation can be fractionated into individual nucleosomes associated with 200 bases of DNA (Fig. 5.6a, track D), dinucleosomes associated with 400 bases of DNA (Fig. 5.6a, track C), and so on (Finch *et al.* 1975). The individual mononucleosomes, dinucleosomes, or larger complexes in each of these fractions can be observed readily in the electron microscope (Fig. 5.6b). These experiments provide direct evidence for the organization of DNA into nucleosomes within the cell.

This organization of DNA into nucleosomes, which can be directly visualized in the electron microscope as the beads on a string structure of chromatin (Fig. 5.7), constitutes the first stage in the packaging of DNA. Subsequently this structure is folded upon itself into a much more compact structure, known as the solenoid (Fig. 5.8). In the formation of this structure histone H1 plays a critical role. As will be seen from Table 5.1, this histone is present at half the level of the other histones and is not part of the core histone octamer. Instead, one molecule of histone H1 seals the two turns which the DNA makes around the core octamer of the other histones. In the solenoid structure, these single histone H1 molecules associate with one another, resulting in tight packing of the individual nucleosomes into a 30 nm fibre (Thoma *et al.* 1979, reviewed by Felsenfeld & McGhee 1986). This 30 nm fibre is the basic structure of chromatin in cells not undergoing division, although during cell division the DNA is further compacted by extensive looping of the 30 nm fibre to form the readily visible chromosomes.

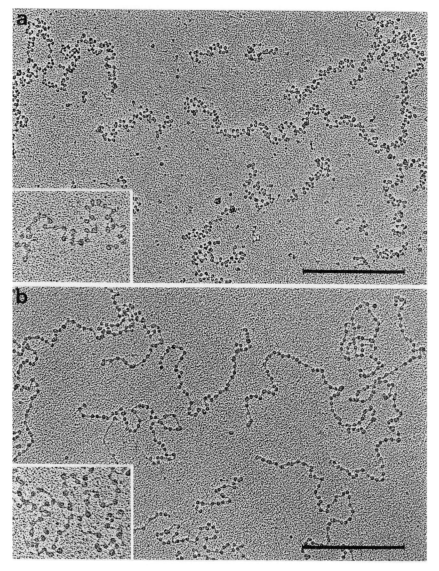

Figure 5.7 Beads on a string structure of chromatin visualized in the electron microscope in the presence (a) or absence (b) of histone H1. The bar indicates 0.5 μm.

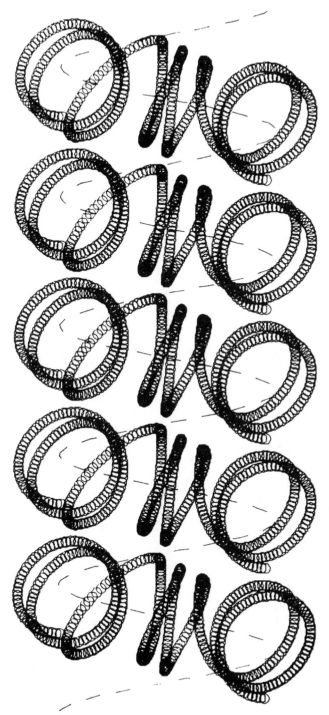

Figure 5.8   The solenoid structure of chromatin.

5.4   CHANGES IN CHROMATIN STRUCTURE IN ACTIVE OR POTENTIALLY ACTIVE GENES

5.4.1   *Active DNA is organized in a nucleosomal structure*

Having established that the bulk of cellular DNA is associated with histone molecules in a nucleosomal structure, an obvious question is whether genes which are either being transcribed or are about to be transcribed in a particular tissue are also organized in this manner or whether they exist as naked, nucleosome-free DNA.

Two main lines of evidence suggest that such genes are organized into nucleosomes. First, if DNA that is being transcribed is examined in the electron microscope, in most cases, the characteristic beads on a string structure (see Section 5.3) is observed, with nucleosomes visible both behind and in front of the RNA polymerase molecules transcribing the gene (Fig. 5.9; McKnight *et al.* 1978). Thus, although this structure may break down in genes which are being extremely actively

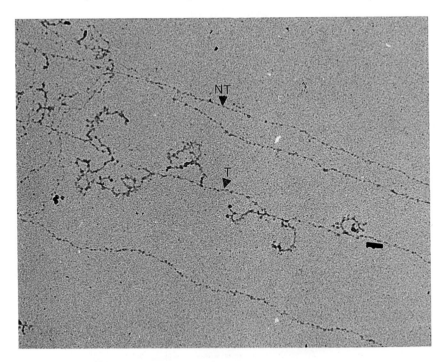

Figure 5.9 Electron micrograph of chromatin from a *Drosophila* embryo. Note the identical 'beads on a string' structure of the chromatin that is not being transcribed (NT) and the chromatin that is being transcribed (T) into the readily visible ribonucleoprotein fibrils.

transcribed, such as occurs for the genes encoding ribosomal RNA during oogenesis, it is maintained in most transcribed genes. Secondly, if DNA organized into nucleosomes is isolated as a ladder of characteristically sized fragments following mild digestion with micrococcal nuclease (see Fig. 5.6), the DNA from active genes is found in these fragments in the same proportion as in total DNA (Lacey & Axel 1975). Similarly, no enrichment or depletion in the amount of a particular gene found in these fragments is observed when the DNA isolated from a tissue actively transcribing the gene is compared to DNA isolated from a tissue that does not transcribe it. The ovalbumin gene, for example, is found in nucleosome-size fragments of DNA in these experiments regardless of whether chromatin from hormonally stimulated oviduct tissue or from liver tissue is used. This is in agreement with the idea that transcribed DNA is still found in a nucleosomal structure rather than as naked DNA, which would be rapidly digested by micrococcal nuclease and hence would not appear in the ladder of nucleosome-sized DNA fragments.

5.4.2 *Sensitivity of active chromatin to DNaseI digestion*

In order to probe for other differences that may exist between transcribed and non-transcribed DNA, many workers have studied the sensitivity of these different regions to digestion with the pancreatic enzyme deoxyribonuclease I (DNaseI; reviewed by Weisbrod 1982, Reeves 1984). Although this enzyme will eventually digest all the DNA in a cell, if it is applied to chromatin in small amounts for a short period, only a small amount of the DNA will be digested. The proportion of transcribed or non-transcribed genes present in the relatively resistant undigested DNA in a given tissue compared to the proportion in total DNA can therefore be used to detect the presence of differences between active and inactive DNA in their sensitivity to digestion with this enzyme. The fate of an individual gene in this procedure can be followed simply by cutting the DNA surviving digestion with an appropriate restriction enzyme and carrying out the standard Southern blotting procedure (Section 2.2.4) with a labelled probe derived from the gene of interest. The presence or absence of the appropriate band derived from the gene in the digested DNA provides a measure of the resistance of the gene to DNaseI digestion (Fig. 5.10).

This procedure has been used to demonstrate that a very wide range of genes that are active in a particular tissue exhibit a heightened sensitivity to DNaseI, which extends over the whole of the transcribed gene and for some distance upstream and downstream of the transcribed region. For example, when chromatin from chick oviduct tissue is digested, the active ovalbumin gene is rapidly digested and its

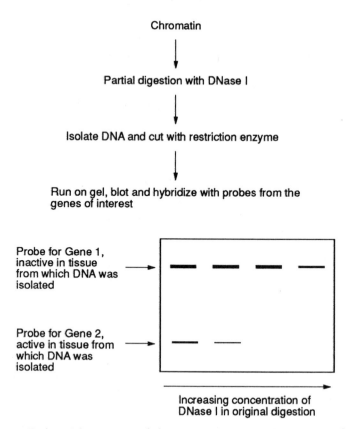

Figure 5.10 Preferential sensitivity of chromatin containing active genes to digestion with DNaseI, as assayed by Southern blotting with specific probes.

characteristic band disappears from the Southern blot under digestion conditions which leave the DNA from the inactive globin gene undigested. This difference between the globin and ovalbumin genes is dependent upon their different activity in oviduct tissue rather than any inherent difference in the resistance of the genes themselves to digestion. Thus, if the digestion is carried out with chromatin isolated from red blood cell precursors in which the globin gene is active and that encoding ovalbumin is inactive, the reverse result to that seen in the oviduct is obtained, with the globin DNA exhibiting preferential sensitivity to digestion and the ovalbumin DNA being resistant to such digestion (Stalder *et al.* 1980a).

It is clear, therefore, that actively transcribed genes, though packaged into nucleosomes, are in a more open chromatin structure than that found for non-transcribed genes, and are hence more accessible to digestion with DNaseI. This altered chromatin structure is not confined to the very active genes such as globin and ovalbumin but appears to be a general characteristic of all transcribed genes, whatever their rate of transcription. Thus, if chromatin is digested using conditions that degrade less than 10 per cent of the total DNA, over 90 per cent of transcriptionally active DNA is digested, the DNA of genes encoding rare mRNAs being as sensitive as that encoding abundant mRNAs. Hence, the altered chromatin structure of active genes does not appear to be dependent upon the act of transcription itself, since the genes encoding rare mRNAs will be transcribed only very rarely.

In agreement with this idea, the altered DNaseI sensitivity of previously active genes persists even after transcription has ceased. For example, the ovalbumin gene remains preferentially sensitive to the enzyme when chromatin is isolated from the oviduct after withdrawal of oestrogen when, as we have previously seen, transcription of the gene ceases (see Ch. 3, Section 3.2.3). A similar preferential sensitivity to DNaseI is also observed for the genes encoding the foetal forms of globin, which are not transcribed in adult sheep cells, and for the adult globin genes in mature (14-day) chicken erythrocytes, following the switching off of transcription.

As well as being detectable after transcription has ceased permanently, such increased sensitivity can also be detected in genes about to become active prior to the onset of transcription. As discussed previously (see Ch. 3, Section 3.2.2), the transcription of the globin gene in Friend erythroleukaemia cells only occurs following treatment of the cells with dimethyl sulphoxide. Increased sensitivity to DNaseI digestion is observed, however, in both the treated cells which transcribe the gene at high levels and in the unstimulated cells which do not (Miller *et al.* 1978).

Hence, the altered, more open, chromatin structure detected by increased DNaseI sensitivity does not reflect the act of transcription itself. Rather, it appears to reflect the ability to be transcribed in a particular tissue or cell type. Hence, in cells that have become committed to a particular lineage expressing particular genes, such commitment will be reflected in an altered chromatin structure which will arise prior to the onset of transcription and will persist after transcription has ceased. The ability of imaginal disc cells in *Drosophila* to maintain their commitment to give rise to a particular cell type in the absence of overt differentiation (see Section 5.2) is likely, therefore, to be due to the genes required in that cell type having already assumed an open chromatin structure. Similarly, the altered chromatin structure of the genes required in cartilage cells would be retained in cells cultured in media not supporting expression of these genes, allowing the restoration of the differentiated cartilage phenotype when the cells are transferred to an appropriate medium (see Section 5.2).

The alteration of chromatin to a more open structure in committed cells is likely to be a necessary prerequisite for gene expression, allowing the *trans*-acting, factors which actually activate the gene, access to the appropriate sequences within it. Hence the different structure of potentially active genes can explain why a particular steroid hormone, such as oestrogen, can induce activity of one particular gene in one tissue and another gene in a different tissue (see Section 5.1). Thus the ovalbumin gene in the oviduct would be in an open configuration, allowing induction to occur, whereas the more closed configuration of the vitellogenin gene would not allow access to the complex of hormone and receptor, and induction would not occur. The reverse situation would apply in liver tissue, allowing induction of the vitellogenin and not the ovalbumin gene.

Therefore, the changes in chromatin structure detected by DNaseI digestion play an important role in establishing the commitment to express the specific genes characteristic of a particular lineage, and it is necessary to investigate the mechanisms responsible for this effect.

5.4.3 Mechanism of increased DNaseI sensitivity

If chromatin is digested with low levels of DNaseI, two low molecular weight proteins are released. These proteins are members of the high-mobility group of non-histone proteins associated with DNA, which gain their name from their small size and high proportion of charged amino acids which result in their moving rapidly in gel electrophoresis. These two proteins, known as HMG 14 and 17, play a crucial role in the preferential sensitivity of active DNA to digestion. Thus, if chromatin is treated with 0.35 M sodium chloride, HMG 14 and 17 are released and the preferential sensitivity to digestion of active DNA is lost. The specific pattern of sensitivity characteristic of the particular tissue can be restored, however, by adding back either or both of these proteins. This suggests that HMG 14 and 17 localize in the region of active or potentially active DNA and produce an altered chromatin configuration in this region.

Given that active DNA is organized into nucleosomes (Section 5.4.1), it seems likely that HMG 14 and 17 function by binding to chromatin and preventing the formation of the more tightly folded solenoidal structure (Section 5.3) in which the nucleosomes are packed more tightly, resulting in reduced accessibility of the DNA to DNaseI. Hence, whereas inactive DNA would exist in the solenoidal structure, active or potentially active DNA would be in the more open beads on a string structure.

Histone H1 is known to be essential for the formation of the solenoid structure and the resulting repression of genes within it (for a review

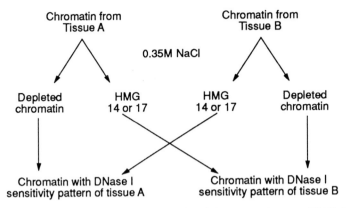

Figure 5.11 Addition of HMG 14 and 17 isolated from one tissue to HMG-depleted chromatin isolated from a second tissue produces the DNaseI sensitivity pattern characteristic of the second tissue.

see Weintraub 1985) and is known to be depleted in active DNA. Hence HMG 14 and 17 may act by displacing histone H1 from potentially active regions, thereby preventing solenoid formation indirectly. Alternatively, their binding to such regions may simply be incompatible with the formation of the solenoid structure, thereby resulting in the more open beads on a string structure.

Whatever the case, the fact that the addition of these proteins to chromatin from which they have been removed restores the preferential sensitivity of active DNA, indicates that these proteins are critical for this effect. This finding also forms the basis for distinguishing whether the tissue specificity of this phenomenon, in which genes are only preferentially sensitive in tissues where they will become active, is caused by these proteins or by other features in the chromatin DNA. It is possible to carry out a mixing experiment (Fig. 5.11) in which the HMG 14 and 17 proteins isolated from one tissue are mixed with the HMG-depleted chromatin of another tissue. If this is done, the pattern of DNaseI sensitivity produced is that of the tissue from which the depleted chromatin was obtained and not that of the tissue from which the HMG proteins were derived. For example, HMG 14 and 17 isolated from brain tissue can restore the preferential sensitivity of the globin gene when added to depleted red blood cell chromatin, whereas HMG 14 and 17 from red blood cells cannot produce this pattern in brain chromatin (Weisbrod & Weintraub 1979).

Hence, although HMG 14 and 17 are critical in producing the open chromatin structure characteristic of active DNA, the tissue-specific differences in the regions of chromatin that are present in this configuration are not paralleled by tissue-specific differences in the nature of HMG 14 and 17. Rather, these proteins are identical in all tissues and must therefore recognize tissue-specific differences in the chromatin DNA, allowing them to bind to particular regions and confer the tissue-specific pattern of DNAseI sensitivity. It is therefore necessary to discuss other features of active chromatin that might be recognized by HMG 14 and 17.

5.5 OTHER CHANGES IN DNA AND ITS ASSOCIATED PROTEINS IN ACTIVE OR POTENTIALLY ACTIVE GENES

5.5.1 DNA methylation

Although DNA consists of the four bases adenine, guanine, cytosine, and thymine, it has been known for many years that these bases can exist in modified forms bearing additional methyl groups. The most common of these in eukaryotic DNA is 5-methyl cytosine (Fig. 5.12), between 2 and 7 per cent of the cytosine in mammalian DNA being modified in this way (reviewed by Razin & Riggs 1980).

Approximately 90 per cent of this methylated C occurs in the dinucleotide, CG, where the methylated C is followed on its 3' side by a G residue. Conveniently, this sequence forms part of the recognition sequence (CCGG) for two restriction enzymes, MspI and HpaII, which differ in their ability to cut at this sequence when the central C is methylated. Thus MspI will cut whether or not the C is methylated and HpaII will only do so if the C is unmethylated. This characteristic allows the use of these enzymes to probe the methylation state of the fraction of CG dinucleotides that is within cleavage sites for these enzymes. Hence, if DNA is digested with either HpaII or MspI, both enzymes will give the same pattern of bands only if all the C residues within the recognition sites are unmethylated. In contrast, if any sites are methylated, larger bands will be obtained in the HpaII digest,

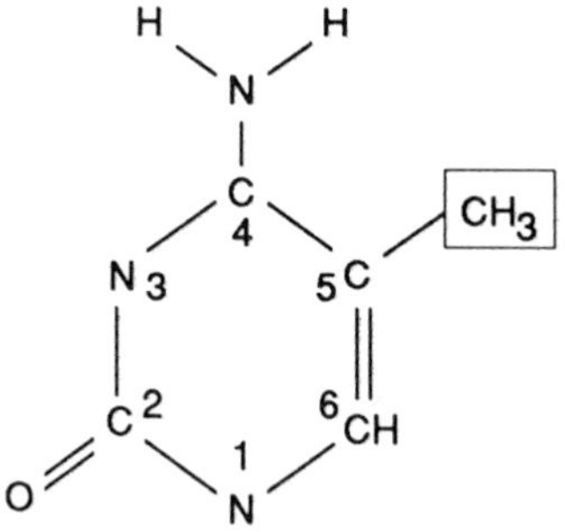

Figure 5.12 Structure of 5-methyl cytosine.

118

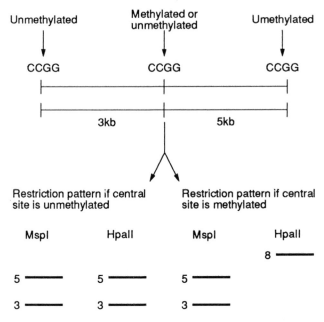

Figure 5.13 Detection of differences in DNA methylation between different tissues, using the restriction enzymes *Msp*I and *Hpa*II.

reflecting the failure of the enzyme to cut at particular sites (Fig. 5.13). If this procedure is used in conjunction with Southern blot hybridization using a probe derived from a particular gene, the methylation pattern of the *Hpa*II/*Msp*I sites within the gene can be determined.

When this is done it is found that although some CG sites are always unmethylated and others are always methylated, a number of sites exhibit a tissue-specific methylation pattern, being methylated in some tissues but not in others. Such sites within a particular gene are unmethylated in tissues where the gene is active or potentially active and methylated in other tissues (for reviews see Doerfler 1983, Cedar 1988). For example, a particular site within the chicken globin gene is methylated in a wide variety of tissues and is therefore not digested with *Hpa*I, but is unmethylated and therefore susceptible to digestion in DNA prepared from erythrocytes (Fig. 5.14). Similarly, the tyrosine amino-transferase gene, which is expressed only in the liver (Section 5.1), is undermethylated in this tissue when compared to other tissues where it is not expressed. The possibility suggested by this finding that *trans*-acting factors, which regulate the expression of this gene and are detectable in all tissues, may be inhibited from binding to the gene in tissues where it is methylated, is confirmed by the observation that

119

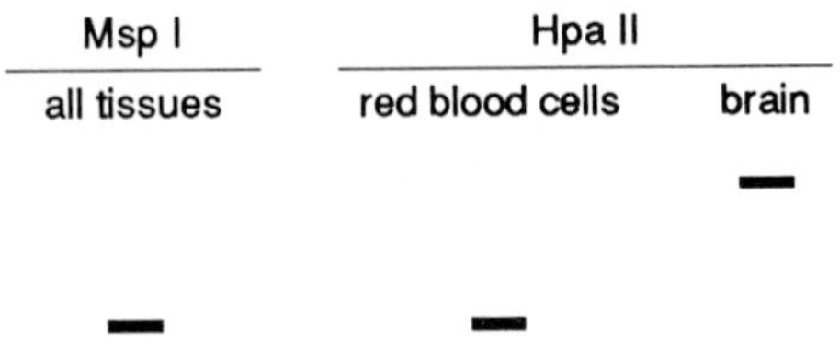

Msp I	Hpa II	
all tissues	red blood cells	brain
		▬
▬	▬	

Figure 5.14 Tissue-specific methylation of *MspI/HpaII* sites in the chicken globin gene results in the methylation-sensitive enzyme, *HpaII*, producing a band in red blood cell DNA that is identical to that produced by the methylation-insensitive *MspI*, but producing a larger band in brain DNA.

artificial methylation of the gene prevents the binding of at least one of these factors (Becker *et al*. 1987).

As with DNaseI sensitivity, in many cases under-methylation is observed prior to the onset of transcription and persists after its cessation. For example, the undermethylation of the chicken globin gene persists in mature erythrocytes, where the gene is not being transcribed but is still sensitive to DNaseI digestion. Most interestingly, the region in which unmethylated C residues are found correlates with that exhibiting heightened DNaseI sensitivity (Weintraub *et al*. 1981) and is also depleted of histone H1 (Ball *et al*. 1983).

As with DNaseI sensitivity, therefore, undermethylation is a consequence of commitment to a particular pattern of gene expression, and represents a possible candidate for the feature in potentially active regions of chromatin which is recognized by HMG 14 and 17, resulting in the more open structure of such regions. That such methylation differences can be recognized by proteins, is shown conclusively both by the fact that the proteinaceous enzyme *HpaII* digests only unmethylated DNA (see above) as well as the finding that the effect of a T to C mutation in the *lac* operon, which prevents binding of the lac repressor protein, can be suppressed by methylating the C residue to yield 5-methyl cytosine.

The idea that methylation differences might be the essential feature causing altered chromatin structure is particularly attractive because of the ease with which such differences can be propagated, allowing cellular commitment to be stable over many generations (see Section 5.1). Thus, in double-stranded DNA, the CG dinucleotide will exist as a symmetrical structure:

$$5' \text{ CG } 3'$$

$$3' \text{ GC } 5'$$

It has been observed that when one C in this structure is methylated,

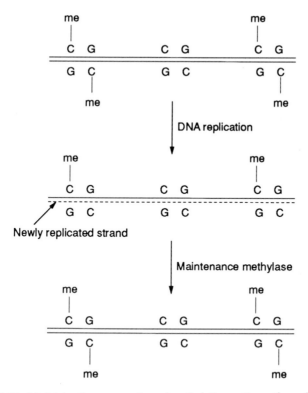

Figure 5.15 Model for the propagation of methylation patterns through cell division.

the C on the opposite strand is also methylated. This effect is achieved by an enzyme present in eukaryotic cells which recognizes sites where only one C is methylated (hemi-methylated sites) and methylates the second C residue rapidly. This enzyme has no activity on a totally unmethylated site, however. Hence, the pattern of DNA methylation will be maintained following DNA replication, with the hemi-methylated sites produced by replication being re-methylated rapidly (Fig. 5.15). Similarly, because the maintenance methylase is only active on hemi-methylated sites, unmethylated sites present in a particular tissue will be propagated through subsequent cell divisions. Thus, once established, a particular pattern of methylation will be maintained, accounting for the stability of the committed state.

Such a mechanism also allows readily for the specific loss of methylation sites that must occur during the process of commitment to a particular lineage. Thus, such losses could occur via a specific demethylation event or simply by inhibiting the action of the maintenance methylase at a particular site following cell division (Fig.

121

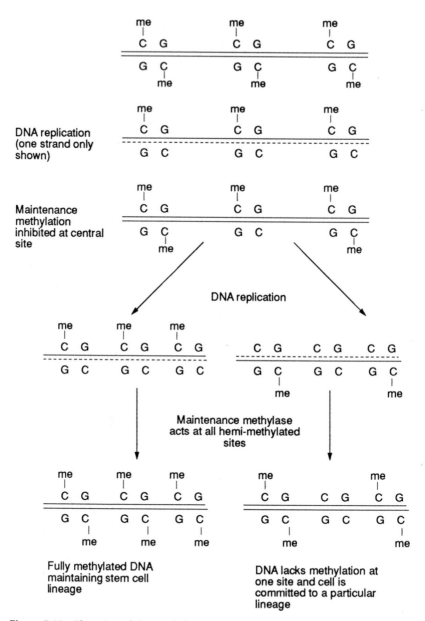

Figure 5.16 Alteration of the methylation pattern to produce an unmethylated site by inhibition of the maintenance methylase and subsequent DNA replication.

5.16). This latter mechanism would eventually result in the generation of one daughter cell at which the site was fully methylated and one in which it was unmethylated. As we have seen (Ch. 2, Sections 2.2.2 & 2.4), such a differentiation event in which a stem cell divides to yield one daughter that differentiates and another that maintains the stem cell lineage is a common feature of embryonic development. As with other methylation patterns, the new demethylated site in the committed cell would be propagated through subsequent cell divisions.

Hence DNA methylation processes provide a means of explaining the stability of the committed state, while allowing for its modification in suitable circumstances. Unlike DNA deletion events, methylation patterns are not irreversible and could be altered when a cell undergoes transdifferentiation or following nuclear transplantation (Ch. 2, Section 2.2.3). As we have discussed, however, such changes in the differentiated state normally require dedifferentiation and cell division, exactly as would be necessary for a process dependent on DNA replication and subsequent inhibition of maintenance methylation at particular sites.

A model in which the regulation of methylation at particular sites controls cellular commitment and the opening of chromatin around active genes is therefore an attractive one. However, the correlation of undermethylation with the ability of a gene to be transcribed is insufficient to prove the truth of this model, and it is necessary to consider other more direct evidence in its favour. This evidence comes from two areas of investigation.

INTRODUCTION OF METHYLATED DNA INTO CELLS

A number of experiments (reviewed by Bird 1984) have shown that if DNA containing 5-methyl cytosine is introduced into cells it is not expressed, whereas the same DNA which has been demethylated is expressed. These experiments have been carried out using both eukaryotic viruses and cellular genes, such as those encoding β- and

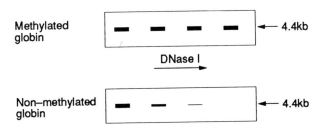

Figure 5.17 Unmethylated DNA introduced into cells adopts an open DNaseI-sensitive configuration, whereas the same DNA when methylated and then introduced into cells adopts a more tightly packed DNaseI-insensitive form.

γ-globin, cloned into plasmid vectors. Most interestingly, in these experiments the methylated DNA adopts a DNaseI-insensitive structure typical of inactive genes whereas unmethylated DNA adopts the DNaseI-sensitive structure typical of active genes (Fig. 5.17; Keshet *et al.* 1986), providing direct evidence for the role of methylation differences in regulating the generation of different forms of chromatin structure.

EFFECT OF ARTIFICIALLY INDUCED DEMETHYLATION

If methylation differences play a crucial role in the regulation of differentiation, it should be possible to change gene expression by demethylating DNA. This has been achieved in a number of cases by treating cells with the cytidine analogue, 5-azacytidine, which is incorporated into DNA but cannot be methylated, having a nitrogen atom instead of a carbon atom at position 5 of the pyrimidine ring (for a review see Cedar 1988). In the most dramatic of these cases, treatment of fibroblasts with this compound results in the activation of a few key regulatory loci and the cells differentiate into multinucleate, twitching, striated muscle cells (Constantinides *et al.* 1977). In other cases, although not actually producing altered gene expression, demethylation may facilitate it. Thus, if HeLa cells are treated with 5-azacytidine no dramatic effects are observed. If such cells are fused with muscle cells, however, muscle-specific genes are switched on in the treated HeLa cells, a phenomenon which is not observed when untreated HeLa cells are fused with mouse muscle cells (Chiu & Blau 1985). Hence, treatment of the HeLa cells has altered their muscle-specific genes in such a way as to allow them to respond to *trans*-acting factors present in the mouse muscle cells. This type of regulatory process is exactly what would be predicted from the role of methylation in the alteration of chromatin structure and thereby in facilitating interactions with *trans*-acting regulatory factors. It should be noted, however, that in none of these cases has it been shown directly that 5-azacytidine achieves its effect by inducing demethylation rather than by some other, as yet uncharacterized, action of this compound.

Despite this caveat, the available evidence from these studies, and those discussed above, indicates that DNA methylation plays a central role in the regulation of gene expression, at least in mammals. It is noteworthy, however, that *Drosophila* does not contain 5-methyl cytosine (Uriel-Shoval *et al.* 1982) and no clear example of a methylated gene has been detected in an invertebrate, although DNaseI sensitivity changes occur in these organisms exactly as in vertebrates. Hence, if methylation of cytosine is indeed the primary cause of changes in chromatin structure in vertebrates, other mechanisms must produce the same effect in invertebrates. Such mechanisms may also be involved in the minority of cases in vertebrates, such as the chicken α 2 (I) collagen gene, where differences in methylation between expressing and non-expressing tissues cannot be detected (McKeon *et al.* 1982). Other features of active chromatin which distinguish it from inactive chromatin and which might be involved in this process will now be discussed.

5.5.2 *Histone modifications*

Given the essential role of histones in chromatin structure, it is possible that potentially active chromatin might be marked in some way by modification of the histones within it. In fact a number of such modifications of these proteins (involving, for example, methylation, phosphorylation, ubiquitination, or acetylation) have been reported. However, we shall discuss only two of these, which have been correlated with active or potentially active regions of chromatin.

UBIQUITINATION

Ubiquitin is a small protein of only 76 amino acids, which forms a conjugate with histone H2A in which the C-terminal carboxyl group of ubiquitin is joined to the free amino group on an internal lysine residue in the histone, to form a branched molecule (Fig. 5.18; reviewed by Busch & Goldknopf 1981). This modification reduces the net positive change on the histone molecule, both by neutralizing the charged amino group on lysine and by introducing a number of negatively charged amino acids present in the ubiquitin molecule itself. Only a small minority of the H2A in a cell (about 5–10 per cent) exists in this form but, in *Drosophila* at least, this modified form of H2A is localized preferentially in nucleosomes containing active genes (Levinger &

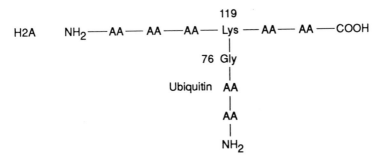

Figure 5.18 Linkage of ubiquitin to histone H2A. The carboxyl terminal amino acid (76) of ubiquitin links to the lysine at position 119 of histone H2A. AA indicates the amino-acid backbone of the molecules.

Varshavsky 1982). Hence, it is possible that ubiquitination may substitute for DNA methylation in regulating chromatin structure in invertebrates, and also in some cases in vertebrates. Unlike DNA methylation, however, the putative role of ubiquitination is based only on the correlation described above, no direct evidence that H2A ubiquitination actually affects gene expression being available.

ACETYLATION

The free amino group on internal lysine residues is also involved in the second modification of histones found in active or potentially active DNA, namely acetylation (for a review see Mathis *et al.* 1980). In this case, however, one of the hydrogen atoms in the free amino group is substituted by an acetyl group ($COCH_3$). This modification, which like ubiquitination reduces the net positive change on the histone molecule, occurs primarily for histones H3 and H4. Hyperacetylated forms of these histones, containing several such acetyl groups, have been shown to be localized preferentially in active genes exhibiting DNaseI sensitivity. Furthermore, treatment of cells with sodium butyrate, which inhibits a cellular deacetylase activity and hence increases histone acetylation, has been shown to result in DNaseI sensitivity of some regions of chromatin and to activate the expression of some previously silent cellular genes (Reeves & Cserjesi 1979). Hence, as with DNA methylation, there is direct evidence that hyperacetylation of histones may play some role in the generation of DNaseI sensitivity and the marking of active or potentially active genes.

It is possible, therefore, that histone modifications such as acetylation or ubiquitination may act as markers, guiding HMG 14 and 17 to certain regions and hence leading indirectly to the destabilization of the solenoid structure. Alternatively, these modifications may affect chromatin structure directly and generate DNaseI sensitivity by altering the charge distribution on the histone molecules, thus affecting the interactions of histones with DNA or each other.

5.6 DNASEI HYPERSENSITIVE SITES IN ACTIVE OR POTENTIALLY ACTIVE GENES

5.6.1 *Detection of DNaseI hypersensitive sites*

So far in this chapter we have seen that the region of chromatin containing an active or potentially active gene has a number of distinguishing features, including undermethylation, histone modifica-

126

tions, and increased sensitivity to digestion with DNaseI. Such changes extend over the entire region of the gene and some flanking sequences and, in the case of DNaseI sensitivity, result in an approximately tenfold increase in the rate at which active or potentially active genes are digested.

Following the discovery of such increased DNaseI sensitivity, many investigators studied whether within the region of increased sensitivity there might be particular sites which were even more sensitive to cutting with the enzyme and which would therefore be cut even before the bulk of active DNA was digested. The technique used to look for such sites is based on that used to look at the overall DNaseI sensitivity of a particular region of DNA (see Section 5.4.2). Chromatin is digested with DNaseI and a restriction enzyme and then subjected to a Southern blotting procedure using a probe derived from the gene of interest. As we have seen previously, the overall sensitivity of the gene can be monitored by observing how rapidly the specific restriction enzyme fragment derived from the gene disappears with increasing amounts of the enzyme (see Fig. 5.10). To search for hypersensitive sites, however, much lower concentrations of the enzyme are used and the appearance of discrete digested fragments derived from the gene is monitored (Fig. 5.19). Such specific fragments have at one end the cutting site for the restriction enzyme used and, at the other, a site at which DNaseI has cut, producing a defined fragment. Since the position at which the restriction enzyme cuts in the gene is known, the position of the hypersensitive site can be mapped simply by determining the size of the fragment produced.

Using this procedure a very wide variety of genes have been shown to contain such hypersensitive sites exhibiting a sensitivity to DNaseI digestion tenfold above that of the remainder of an active gene and therefore about one hundredfold above that seen in inactive DNA (for a review see Gross & Garrard 1988). A representative list of cases in which such sites have been detected is given in Table 5.2. As with the increased sensitivity of the gene itself, many hypersensitive sites appear only in tissues where the gene is active. Thus, the increased sensitivity of globin DNA in erythrocytes to digestion is paralleled by the presence of hypersensitive sites within the gene in erythrocyte chromatin but not in that of other tissues (Stalder et al. 1980b). Similarly, the ovalbumin gene in hormonally treated chick oviduct also exhibits hypersensitive sites that are not found in other tissues, including erythrocytes (Kaye et al. 1984; Fig. 5.20).

As with undermethylation and the sensitivity of the entire gene to digestion, DNaseI hypersensitive sites appear to be related to the potential for gene expression rather than always being associated with the act of transcription itself. Thus the hypersensitive sites near the

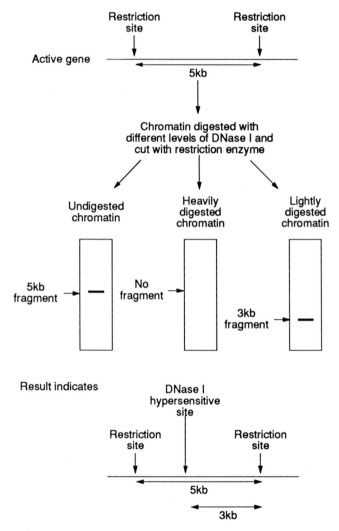

Figure 5.19 Detection of DNaseI hypersensitive sites in active genes by mild digestion of chromatin to produce a digestion product with a restriction site at one end and a DNaseI hypersensitive site at the other (right-hand panel). More extensive digestion will result in the disappearance of the band (central panel) as in the experiment illustrated in Figure 5.10.

Drosophila heat-shock genes are present in the chromatin of embryonic cells prior to any heat-induced transcription of these genes (Keene *et al.* 1981), and one of the sites in the mouse α-foetoprotein gene persists in the chromatin of adult liver after the transcription of the gene (which is confined to the foetal liver) has ceased (Nahon *et al.* 1987).

Hence, as with DNA methylation and generally increased sensitivity

Table 5.2

Examples of genes containing DNase I hypersensitive sites

a) Tissue specific genes

Immune system:–	Immunoglobulin,complement C4
Red blood cells:–	Alpha, beta and epsilon globin
Liver:–	Alpha foetoprotein, serum albumin
Nervous system:–	Acetylcholine receptor
Pancreas:–	Preproinsulin, elastase
Connective tissue:–	Collagen
Pituitary gland:–	Prolactin
Salivary gland:–	*Drosophila* glue proteins
Silk gland:–	Silk moth fibroin

b)Inducible genes

Steroid hormones:–	Ovalbumin, vitellogenin, tyrosine amino-transferase
Stress:–	Heat shock proteins
Viral infection:–	Beta interferon
Amino acid starvation:–	Yeast HIS 3 gene
Carbon source:–	Yeast GAL genes, yeast ADH II gene

c) Others

Histones, ribosomal RNA, 5S RNA, transfer RNA, cellular oncogenes c-*myc* and c-*ras*, glucose-6-phosphate dehydrogenase, dihydrofolate reductase, cysteine protease etc.

to DNaseI, the appearance of hypersensitive sites appears to be involved in gene regulation. This idea is reinforced by the location of the hypersensitive sites, which can be precisely mapped as described above. Many sites are located at the 5′ end of the genes, in positions corresponding to DNA sequences that are known to be important in regulating transcription. For example, a site present at the 5′ end of the steroid-inducible tyrosine amino-transferase gene is localized within the DNA sequence that is responsible for the steroid inducibility of the gene (see Ch. 6, Section 6.2.3). Even in cases where hypersensitive

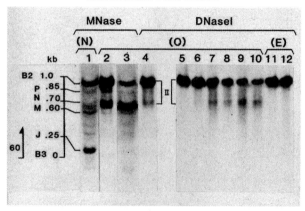

Figure 5.20 Detection of a DNaseI hypersensitive site in the ovalbumin gene in oviduct tissue (0) but not in erythrocytes (E). Track 4 shows the detection of a lower band caused by cleavage at a hypersensitive site when oviduct chromatin is digested with DNaseI. *Note the progressive appearance of this band as increasing amounts of DNaseI are used to digest the oviduct chromatin (tracks 5 to 10). No cleavage is observed when similar amounts of DNaseI are used to cut erythrocyte chromatin (tracks 11 and 12). The hypersensitive site in oviduct chromatin is also cleaved, however, with micrococcal nuclease (tracks 2 to 3). Track 1 shows the pattern produced by micrococcal nuclease cleavage of naked DNA (N).

sites are located far from the site of transcriptional initiation, they appear to correspond to other regulatory sequences, such as enhancers, which can act over large distances (see Ch. 6, Section 6.3).

In the case of the *Drosophila* gene encoding the glue protein Sgs4, the fortuitous existence of a mutant strain of fly has indicated the functional importance of hypersensitive sites. In normal flies this gene contains two hypersensitive sites, 405 and 480 bases upstream of the start of transcription. In the mutant, both sites are removed by a small DNA deletion of 100 base pairs. Despite the fact that this gene still has the start site of transcription and 350 bases of upstream sequences, no transcription occurs (Fig. 5.21), indicating the regulatory importance of

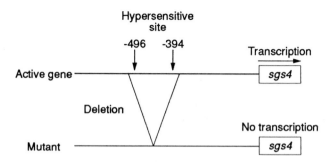

Figure 5.21 Deletion of a region containing the two hypersensitive sites upstream of the *Drosophila* sgs4 gene abolishes transcription.

the region containing the hypersensitive sites (Shermoen & Beckendorf 1982).

It is clear, therefore, that hypersensitive sites represent another marker for active or potentially active chromatin and are a feature likely to be of particular importance in gene regulation, being associated with many DNA sequences that regulate gene expression. It is therefore necessary to consider the nature and significance of these sites.

5.6.2 *Nature and significance of hypersensitive sites*

In many cases DNaseI hypersensitive sites are also exquisitely sensitive to cleavage with other enzymes, such as the S1, Bal31, and micrococcal nucleases (Fig. 5.20; reviewed by Elgin 1984), which are particularly active on DNA that is single stranded or has an altered configuration. Although it is unlikely that the DNA in hypersensitive sites is actually single stranded, since it is also cut by enzymes specific for double-stranded DNA, it is likely that it exists in a configuration different from that of the bulk DNA. This may involve, for example, the DNA in this region being in a highly supercoiled form which is under torsional stress. Such supercoiling has been shown to stabilize an alternative structural form of DNA, known as Z-DNA, in which the DNA helix coils in a left-handed rather than the conventional right-handed manner (for a review see Rich *et al.* 1984).

A variety of evidence suggests that the formation of Z-DNA is associated with transcriptional activity. Thus antibodies that specifically recognize Z-DNA are known to stain the heat-inducible puffs in *Drosophila* polytene chromosomes (Pardue *et al.* 1983), as well as the transcriptionally active macronucleus of the ciliated protozoan *Stylonichia*. Similarly, sites that form Z-DNA (consisting of alternating purines and pyrimidines) preferentially are associated closely with DNaseI hypersensitive sites in a number of different situations, for example within the enhancer element that regulates transcription of the eukaryotic virus SV40 (Fig. 5.22).

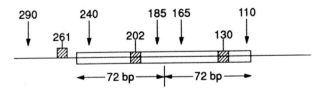

Figure 5.22 Association of hypersensitive sites (arrows) with regions of Z-DNA (shaded) in the DNA of the SV40 enhancer element. The boxed regions indicate the 72 base pair sequence, which is repeated twice in the enhancer. The numbers indicate the position of each element within the enhancer.

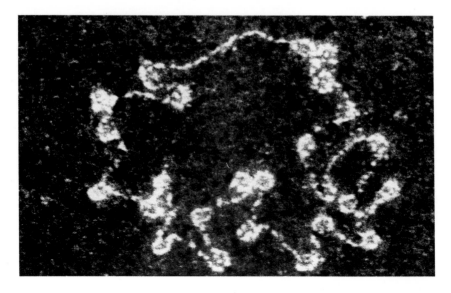

Figure 5.23 Electron micrograph of the SV40 mini-chromosome consisting of DNA and associated histones. Note the region of the enhancer and hypersensitive sites, which appears as a thin filament of DNA free of associated proteins.

When this virus enters cells its DNA, which is circular and only 5000 bases in size, becomes associated with histones in a typical nucleosomal structure, which can be visualized in the electron microscope as a mini-chromosome (Saragosti *et al*. 1980; Fig. 5.23). When this is done, however, the region containing the hypersensitive sites and potential Z-DNA-forming regions remains nucleosome free and is seen as naked DNA. A similar lack of nucleosomes in the region of hypersensitive sites is also found in the chicken β-globin gene, the 5' hypersensitive site of this gene being excisable as a 115 bp restriction fragment lacking any associated nucleosomes (McGhee *et al*. 1981).

It is clear, therefore, that hypersensitive sites represent regions of DNA which are free of nucleosomes and which may have an unusual configuration. Such a non-nucleosomal structure is likely to facilitate the entry of regulatory proteins or RNA polymerase itself into the DNA and therefore to allow the onset of transcription. In agreement with this idea, studies on the interaction of RNA polymerase with chromatin in the test-tube have shown that the enzyme cannot initiate transcription on DNA that is packaged in nucleosomes, although as previously discussed (Section 5.4.1) it can elongate a previously initiated transcript through a region that is packaged in this way (Lorch *et al*. 1987).

Hence, the existence of a nucleosome-free site is probably essential

a) Heat shock genes

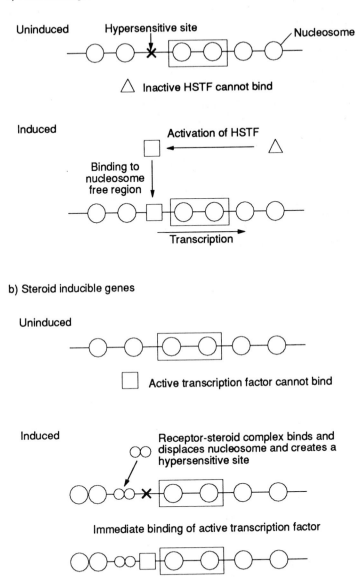

Figure 5.24 Two mechanisms for transcriptional activation. (a) The heat shock transcription factor is activated by heat and binds to a pre-existing nucleosome-free region. (b) The receptor–steroid complex displaces a nucleosome, creating a hypersensitive site, and allowing an active transcription factor to bind. X indicates the position of a hypersensitive site.

for the onset of transcription. As with the general sensitivity of active genes to DNaseI or undermethylation, the existence of such a site is likely to be necessary but not sufficient for transcription, which will require the binding to the DNA of other *trans*-acting factors or of RNA polymerase itself. In the case of the heat-shock genes whose transcription is stimulated by elevated temperatures (see Ch. 3, Section 3.2.4) hypersensitive sites are present prior to heat treatment. Following heat treatment, a protein factor, known as the heat-shock transcription factor (HSTF), binds to this region of DNA and transcription begins (see Ch. 6, Section 6.2.2). In this case, the HSTF is only capable of binding following heat shock and hence transcription only occurs following such treatment (Fig. 5.24a). In other cases, however, where the necessary transcription factors are present in all tissues, transcription may follow immediately the nucleosome-free region is generated, allowing these factors access. Thus, in the case of glucocorticoid-responsive genes (reviewed by Beato 1989), the critical regulatory event is the binding of the glucocorticoid receptor/steroid complex to a particular DNA sequence, which displaces a nucleosome and generates a DNaseI hypersensitive site. Ubiquitous factors present in all tissues, such as NFI and the TATA box binding factor, immediately bind to this region and transcription begins (Fig, 5.24b).

Although these two situations appear different in terms of the time at which the hypersensitive site appears relative to the onset of transcription, they illustrate the basic role of hypersensitive sites, namely the generation of a site of access for regulatory proteins.

5.7   CONCLUSIONS

A variety of changes take place in the chromatin of genes during the process of commitment to a particular pathway of differentiation. Such changes involve both modification of the DNA itself by undermethylation, to the histones with which it is associated, and to the general packaging of the DNA in chromatin. They result in three levels of chromatin structure within the cell (Fig. 5.25). Thus although the bulk of inactive DNA is organized into a tightly packed solenoid structure, active or potentially active genes are organized into a more open 'beads on a string' structure, and short regions within the gene are present in entirely nucleosome-free DNA which may have an unusual conformation.

The role of these changes in allowing cells to maintain a commitment to a particular differentiated state and to respond differently to inducers of gene expression is well illustrated in the case of the steroid hormones and their effect on gene expression. Thus the difference in

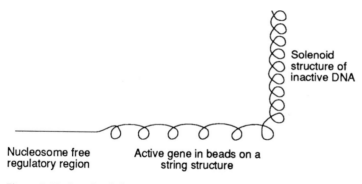

Figure 5.25 Levels of chromatin structure in active and inactive DNA.

the response of different tissues to treatment with oestrogen (see Section 5.1) is likely to be due to the fact that in one tissue certain steroid-responsive genes will be inaccessible within the solenoid structure and will therefore be incapable of binding the receptor–hormone complex that is necessary for activation. In other genes, which are in the beads on a string structure and therefore more accessible, such binding of the complex to defined sequences in the gene will occur. Even in this case, however, gene activation will not occur as a direct consequence of this interaction as might be the case in bacteria. Rather, the binding will result in the displacement of a nucleosome from the region of DNA, generating a hypersensitive site and allowing other regulatory proteins to interact with their specific recognition sequences and cause transcription to occur.

Both the action of the receptor–hormone complex and the subsequent binding of other transcription factors to nucleosome-free DNA clearly involve the interaction of regulatory proteins with specific DNA sequences. The next two chapters will discuss the DNA sequences and proteins involved in these interactions.

REFERENCES

Ball, D. J., D. S. Gross & W. T. Garrard 1983. 5-methyl cytosine is localized in nucleosomes that contain H1. *Proceedings of the National Academy of Sciences of the USA* **80**, 5490–4.

Beato, M. 1989. Gene regulation by steroid hormones. *Cell* **56**, 335–44.

Becker, P. B., S. Ruppert & G. Schutz 1987. Genomic fingerprinting reveals cell type-specific binding of ubiquitous factors. *Cell* **51**, 435–43.

Bird, A. P. 1984. DNA methylation – how important in gene control? *Nature* **307**, 503.

Busch, H. & I. L. Goldknopf 1981. Ubiquitin–protein conjugates. *Molecular and Cellular Biochemistry* **40**, 173–87.

Cedar, H. 1988. DNA methylation and gene activity. *Cell* **53**, 3–4.

Chiu, C.-P. & H. Blau 1985. 5-azacytidine permits gene activation in a previously non-inducible cell type. *Cell* **40**, 417–24.

Constantinides, P. G., P. A. Jones & W. Gevers 1977. Functional striated muscle cells from non-myoblast precursors following 5-azacytidine treatment. *Nature* **267**, 364–6.

Coon, H. G. 1966. Clonal stability and phenotypic expression of chick cartilage cells *in vitro*. *Proceedings of the National Academy of Sciences of the USA* **55**, 6673.

Doerfler, W. C. 1983. DNA methylation and gene activity. *Annual Review of Biochemistry* **52**, 93–124.

Elgin, S. C. R. 1984. Anatomy of hypersensitive sites. *Nature* **309**, 213–14.

Felsenfeld, G. & J. D. McGhee 1986. Structure of the 30 nm chromatin fiber. *Cell* **44**, 375–7.

Finch, J. T., M. Noll & R. D. Kornberg 1975. Electron microscopy of defined lengths of chromatin. *Proceedings of the National Academy of Sciences of the USA* **72**, 3320–2.

Gottesman, S. 1984. Bacterial regulation: Global regulatory networks. *Annual Review of Genetics* **18**, 415–41.

Gross, D. S. & W. T. Garrard 1988. Nuclease hypersensitive sites in chromatin. *Annual Review of Biochemistry* **57**, 159–97.

Hadorn, E. 1968. Transdetermination in cells. *Scientific American* **219** (Nov.) 110–20.

Igo-Kemenes, T., W. Horz & H. G. Zachau 1982. Chromatin. *Annual Review of Biochemistry* **51**, 89–121.

Kaye, J. S., M. Bellard, G. Dretzen, F. Bellard & P. Chambon 1984. A close association between sites of DNase I hypersensitivity and sites of enhanced cleavage by microccocal nuclease in the 5' flanking region of the actively transcribed ovalbumin gene. *EMBO Journal* **3**, 1137–44.

Keene, M. A., V. Corces, K. Lowenhaupt & S. Elgin 1981. DNase I hypersensitive sites in *Drosophila* chromatin occur at the 5' end of regions of transcription. *Proceedings of the National Academy of Sciences of the USA* **78**, 143–6.

Koshet, I., J. Lieman-Hurwitz & H. Cedar 1986. DNA methylation affects the formation of active chromatin. *Cell* **44**, 535–44.

Kornberg, R. D. & A. Klug 1981. The nucleosome. *Scientific American* **244**, (Feb.), 48–64.

Lacey, E. & R. Axel 1975. Analysis of DNA of isolated chromatin subunits. *Proceedings of the National Academy of Sciences of the USA* **72**, 3978–82.

Levinger, L. & A. Varshavsky 1982. Selective arrangement of ubiquitinated and D1 protein containing nucleosomes within the *Drosophila* genome. *Cell* **28**, 375–85.

Lewin, B. 1980. *Gene expression*. Vol. 2: *Eukaryotic chromosomes*. New York: Wiley.

Lorch, Y., J. W. La Pointe & R. D. Kornberg 1987. Nucleosomes inhibit the initiation of transcription but allow chain elongation with the displacement of histones. *Cell* **49**, 203–10.

McGhee, J. D., W. I. Wood, M. Dolan, J. D. Engel & G. Felsenfeld 1981. A 200 base pair region at the 5' end of the chicken adult β-globin gene is accessible to nuclease digestion. *Cell* **27**, 45–55.

McKeon, C., H. Ohkubo, I. Pastan & B. de Crombrugghe 1982. Unusual methylation pattern of the alpha 2 (1) collagen gene. *Cell* **29**, 203–10.

McKnight, S. L., M. Bustin & O. L. Miller 1978. Electron microscope analysis of chromosome metabolism in the *Drosophila melanogaster* embryo. *Cold Spring Harbor Symposium on Quantitative Biology* **42**, 741–54.

Mathis, D., P. Oudet & P. Chambon 1980. Structure of transcribing chromatin. *Progress in Nucleic Acids Research and Molecular Biology* **24**, 1–55.

Miller, D. M., P. Turner, A. W. Nienhuis, D. E. Axelrod & T. U. Gopalakrishnan 1978. Active conformation of the globin genes in uninduced and induced mouse erythroleukemia cells. *Cell* **14**, 511–21.

Miller, J. & W. K. Reznikoff, (eds) 1980. *The operon*. New York: Cold Spring Harbor Laboratory.

Morse, R. H. & R. T. Simpson 1988. DNA in the nucleosome. *Cell* **54**, 285–7.

Nahon, J.-L, A. Venetianer & J. M. Sala-Trepat 1987. Specific sets of DNase I hypersensitive sites are associated with the potential and overt expression of the rat albumin and alpha-fetoprotein genes. *Proceedings of the National Academy of Sciences of the USA* **84**, 2135–9.

Pardue, M. L., A. Nordheim, E. M. Lafer, B. D. Stollar & A. Rich 1983. Z-DNA and the polytene chromosome. *Cold Spring Harbor Symposia on Quantitative Biology* **47**, 171–6.

Raibaud, O. & M. Schwartz 1984. Positive control of transcription initiation in bacteria. *Annual Review of Genetics* **18**, 173–206.

Razin, A. & A. D. Riggs 1980. DNA methylation and gene function. *Science* **210**, 604–10.

Reeves, R. 1984. Transcriptionally active chromatin. *Biochimica et Biophysica Acta* **782**, 343–93.

Reeves, R. & P. Cserjesi 1979. Sodium butyrate induces new gene expression in Friend erythroleukemic cells. *Journal of Biological Chemistry* **254**, 4283–90.

Rich, A., A. Nordheim & A. H.-J. Wang 1984. The chemistry and biology of left-handed DNA. *Annual Review of Biochemistry* **53**, 791–846.

Saragosti, S., G. M. Moyne & M. Yaniv 1980. Absence of nucleosomes in a fraction of SV40 chromatin between the origin of replication and the region coding for the late leader RNA. *Cell* **20**, 65–73.

Shermoen, A. W. & S. K. Beckendorf 1982. A complex of interacting DNAse I hypersensitive sites near the *Drosophila* glue protein gene sgs4. *Cell* **29**, 601–7.

Stalder, J., M. Groudine, J. B. Dodgson, J. D. Engel & H. Weintraub 1980a. Hb switching in chickens. *Cell* **19**, 973–80.

Stalder, J., A. Larsen, J. D. Engel, M. Dolan, M. Groudine & H. Weintraub 1980b. Tissue-specific DNA cleavages in the globin chromatin domain introduced by DNAse I. *Cell* **20**, 451–60.

Thoma, F., T. Koller & A. Klug 1979. Involvement of histone H1 in the organization of the nucleosome and of the salt-dependent superstructures of chromatin. *Journal of Cell Biology* **83**, 403–27.

Uriel-Shoval, S., Y. Grunebaum, J. Sedat & A. Razin 1982. The absence of detectable methylated bases in *Drosophila melanogaster* DNA. *FEBS Letters* **146**, 148–52.

Weintraub, H. 1985. Assembly and propagation of repressed and derepressed chromatin states. *Cell* **42**, 705–11.

Weintraub, H., A. Larsen & M. Groudine 1981. Alpha globin gene switching during the development of chicken embryos: expression and chromosome structure. *Cell* **24**, 333–44.

Weisbrod, S. 1982. Active chromatin. *Nature* **297**, 289–95.

Weisbrod, S. & H. Weintraub 1979. Isolation of a sub-class of nuclear proteins responsible for conferring a DNAse I-sensitive structure on globin chromatin. *Proceedings of the National Academy of Sciences of the USA* **76**, 630–4.

CHAPTER SIX

Transcriptional control – DNA sequence elements

6.1 INTRODUCTION

6.1.1 *Relationship of gene regulation in prokaryotes and eukaryotes*

As discussed in Chapter 5, various alterations occur in the chromatin structure of a particular gene prior to the onset of transcription. Once such changes have occurred, the actual onset of transcription takes place through the interaction of defined proteins (transcription factors) with specific DNA sequences adjacent to the gene. This final stage of gene regulation is clearly analogous to the activation or repression of gene expression in prokaryotes, which was discussed briefly in Chapter 5 (Section 5.1). However, before the manner in which transcription factors and DNA sequences act to regulate gene expression in higher eukaryotes can be considered, it is necessary to mention two features of eukaryotic systems that do not exist in bacteria.

6.1.2 *Complexity of the eukaryotic system*

RNA POLYMERASES

In prokaryotes a single RNA polymerase enzyme is responsible for the transcription of DNA into RNA. In eukaryotes this is not the case, and three such enzymes, active on distinct sets of genes, exist and can be distinguished by their relative sensitivity to the fungal toxin α-amanitin (Table 6.1). Thus, whereas all genes capable of encoding a protein, as well as the genes for some small nuclear RNAs involved in RNA splicing (see Ch. 4, Section 4.2), are transcribed by RNA polymerase II, the genes encoding the 28S, 18S, and 5.8S ribosomal RNAs are transcribed by RNA polymerase I, and those encoding the transfer RNAs and the 5S ribosomal RNA are transcribed by RNA polymerase III (reviewed by Lewis & Burgess 1982).

138

Table 6.1

Eukaryotic RNA polymerases

	Genes transcribed	Sensitivity to alpha-amanitin
I	Ribosomal RNA (45S precursor of 28S, 18S and 5.8S rRNA)	Insensitive
II	All protein coding genes, small nuclear RNAs U1, U2, U3 etc	Very sensitive (inhibited 1μg/ml)
III	Transfer RNA, 5.8S ribosomal RNA, small nuclear RNA U6, repeated DNA sequences:– Alu, B1, B2 etc.,7SK, 7SL RNA.	Moderately sensitive (inhibited 10μg/ml)

In considering the transcriptional regulatory processes that produce tissue-specific variation in mRNA and protein, our primary concern will therefore be with the regulation of RNA polymerase II transcription. The DNA sequence elements involved in regulating transcription by this enzyme are considered in Sections 6.2–6.4 of this chapter, and the proteins that bind to them are discussed in Chapter 7. Transcription by RNA polymerases I and III is also subject to regulation, however, and has been analysed in particular detail in the case of the transcription of the genes encoding the oocyte-specific form of 5S ribosomal RNA by RNA polymerase III. The processes regulating transcription by these polymerases are distinct from those operating on RNA polymerase II, and are considered in a separate section of this chapter (Section 6.5).

CO-ORDINATELY REGULATED GENES ARE NOT LINKED IN EUKARYOTES

Even when considering transcription by RNA polymerase II alone, eukaryotic genes exhibit one further complication when compared to prokaryotes. In prokaryotes, when several genes encoding particular proteins are expressed in response to a particular signal, the genes are found tightly linked together in an operon. In response to the activating signal, all the genes in the operon are transcribed as one single polycistronic (multi-gene) mRNA molecule, and translation of this molecule results in the desired co-ordinate production of the proteins encoded by the individual genes. A typical example of this is seen in the case of the lactose-inducible genes located in the *lac* operon. The efficient use of lactose requires not only the enzyme β-galactosidase, which cleaves the lactose molecule, but also a permease enzyme, to facilitate uptake of lactose by the cell, and a modifying *trans*-acetylase enzyme. The genes encoding these three molecules must be activated in response to lactose and this is achieved by linking them together in

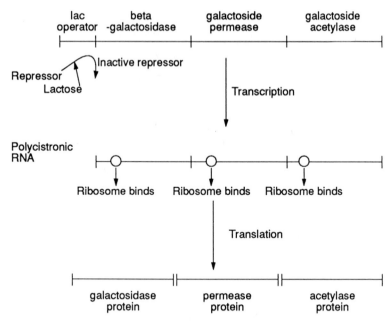

Figure 6.1   Structure of the *lac* operon of *E. coli*, in which the three genes are transcribed into one single RNA and translated into individual proteins following binding of the ribosome at three sites in the RNA molecule.

an operon whose expression is dependent upon the presence of lactose (Fig. 6.1).

In higher eukaryotic organisms this does not occur. Thus operon-type structures transcribed into polycistronic RNAs do not exist in higher organisms. Rather, individual genes are transcribed into individual monocistronic RNAs encoding single proteins. Hence, genes whose protein products are required in parallel are present in the genome as individual genes whose expression must be co-ordinated. Moreover, such co-ordinately expressed genes are not even closely associated in the genome in a manner that might be thought to facilitate their regulation, but are very often present on different chromosomes within the eukaryotic nucleus. Thus, the production of a functional antibody molecule by the mammalian B-cell requires the synthesis of both the immunoglobulin heavy- and light-chain proteins, which together make up the functional antibody molecule. The genes encoding the heavy- and light-chain proteins are, however, found on separate chromosomes; the heavy-chain locus being on chromosome 14 in humans whereas light-chain genes are found on both chromosomes 2 and 22. Similarly, the production of a functional globin molecule requires the association of an α-globin-type protein and a β-globin-type protein. Yet the genes encoding the various members of the α-globin family are on chromosome 16 in humans whereas the genes encoding the β-globins are found on chromosome 11 (for a review of the globin gene family see Maniatis *et al.* 1980).

This difference between prokaryotes and eukaryotes is likely to reflect a need for greater flexibility in regulating gene expression in eukaryotes. Thus the bacterial arrangement, although providing a simple mechanism for co-ordinating the expression of different proteins, also means that, in general, expression of one protein will also necessitate expression of the other co-ordinately expressed proteins. In contrast, the eukaryotic system allows α-globin to be produced in parallel with β-globin in the adult organism but to be associated with another β-globin-like protein, namely γ-globin, in the foetus.

6.1.3   *The Britten and Davidson model for the co-ordinate regulation of unlinked genes*

This greater flexibility does, however, necessitate some means of co-ordinately regulating gene expression, allowing the production of α- and β-globin in the adult reticulocyte, α- and γ-globin in the embryonic reticulocyte, and the heavy and light chains of immunoglobulin in the antibody-producing B-cell. A model of the mechanism of such co-ordinate regulation was put forward by Britten & Davidson (1969). They proposed that genes regulated in parallel with one another in response to a particular signal would contain a common regulatory element which would cause the activation of the gene in response to that signal (Fig. 6.2). Individual genes could contain more than one regulatory element, some of which would be shared with other genes, which in turn could possess elements not present in the first gene. Specific signals causing gene activation would act by stimulating a specific integrator gene whose product would activate all the genes containing one particular sequence element. This mechanism would allow the observed activation of distinct but overlapping sets of genes in response to specific signals via the activation of particular integrator genes.

Although this model was proposed over 20 years ago, when our understanding of eukaryotic gene regulation was considerably more limited than at present, it continues to serve as a useful framework for considering gene regulation. In modern terms, the integrator gene would be considered as encoding a transcription factor (see Ch. 7) which binds to the regulatory sequences and activates expression of the corresponding genes. As discussed in Chapter 7 (Section 7.4), the activation of such a factor by a particular signal has now been shown to

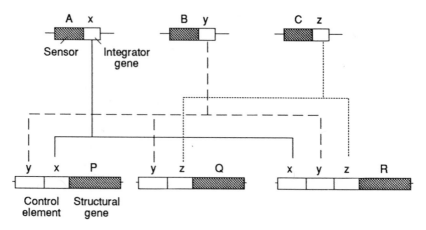

Figure 6.2 Britten and Davidson model for the co-ordinate expression of unlinked genes in eukaryotes. Sensor elements (A, B, C) detect changes requiring alterations in gene expression and switch on appropriate integrator genes (x, y, z) whose products activate the structural genes (P, Q, R) containing the appropriate control elements. Note the flexibility of the system whereby a particular structural gene can be activated with or without another structural gene by selecting which integrator gene is activated.

operate both by *de novo* synthesis of the factor (in an individual tissue or in response to a particular signal) as envisaged by Britten and Davidson, and by direct activation of a pre-existing protein, for example, by phosphorylation in response to the activating signal. The concept that such factors act by binding to DNA sequences held in common between co-ordinately regulated genes is now well established, and the nature of such sequences will be considered in Sections 6.2–6.4 (for a recent review of this topic see Maniatis *et al.* 1987).

6.2   SHORT SEQUENCE ELEMENTS LOCATED WITHIN OR ADJACENT TO THE GENE PROMOTER

6.2.1   *Short regulatory elements*

In prokaryotes the sequence elements that play an essential role in transcriptional regulation are located immediately upstream of the point at which transcription begins, and form part of the promoter which directs transcription of the gene. Such sequences can be divided into two classes, namely those found in all genes, which play an essential role in the process of transcription itself, and those found in one or a few genes, which mediate their response to a particular signal (Schmitz & Galas 1979). It would be expected by analogy, therefore, that the sequences that play a role in the regulation of eukaryotic transcription would be located similarly, upstream of the transcription start site within or adjacent to the promoter. Hence, a comparison of such sequences in different genes should reveal both basic promoter elements necessary for all transcription (which would be present in all genes) and those necessary for a particular pattern of regulation (which would be present only in genes exhibiting a specific pattern of regulation). The role of such sequences can be confirmed either by destroying them by deletion or mutation or by transferring them to other genes in an attempt to confer the specific pattern of regulation of the donor gene upon the recipient.

6.2.2   The heat-shock response element

To illustrate this method of analysis, we will focus upon the gene encoding the 70 KDa molecular weight heat-shock protein (hsp70). As discussed in Chapter 3 (Section 3.2.4), exposure of a very wide variety of cells to elevated temperature results in the increased synthesis of a few heat-shock proteins, of which hsp70 is the most abundant. Such increased synthesis is mediated in part by increased transcription of the corresponding gene, which can be visualized as a puff within the polytene chromosome of *Drosophila* (for a review see Ashburner & Bonner 1979). Hence, examination of the sequences located upstream of the start site for transcription in this gene should identify potential sequences involved in its induction by temperature elevation, as well as those involved in the general mechanism of transcription. The sequences present in this region of the *hsp70* gene which are also found in other genes are listed in Table 6.2, and their arrangement is illustrated in Figure 6.3 (Williams *et al.* 1989; for further information on these sequences see Davidson *et al.* 1983, Jones *et al.* 1988 and references therein).

   A comparison of this type reveals a number of sequence motifs shared by the *hsp70* gene and other non-heat-inducible genes. Several of these are involved in the general process of transcription. The TATA box, for example, is found approximately 30 bases upstream of the transcription start site in a wide variety of different genes, although it is absent from some housekeeping genes expressed in all tissues and a few tissue-specific genes. In genes that contain it, the TATA box plays a critical role in positioning the start site of transcription, and its destruction by mutation or deletion effectively abolishes transcription of such genes (for a review see Breathnach & Chambon 1981). Similarly, the CCAAT box, which is located farther upstream of the start site of transcription of a wide variety of genes (including the *hsp70* gene) which are regulated in different ways, is also believed to play an

Table 6.2

Sequences present In the up-stream region of the hsp70 gene which are slso found in other genes

Name	Consensus	Other genes containing sequences
TATA box	TATA A/T A A/T	Very many genes
CCAAT box	TGTGGCTNNNAGCCAA	Alpha and beta globin, herpes simplex virus thymidine kinase, cellular oncogenes c-*ras*, c-*myc*, albumin etc
Sp1 box	GGGCGG	Metallothionein II A, type II procollagen, dihydrofolate reductase etc
CRE	T/G T/A CGTCA	Somatostatin, fibronectin, alpha-gonadotrophin, c-*fos* etc
AP2 box	CCCCAGGC	Collagenase, class 1 antigen H-2K^b, metallothionein II A
Heat shock consensus	CTNGAATNTTCTAGA	Heat inducible genes hsp83, hsp27 etc

important role in allowing transcription of the genes containing it (for a review see McKnight & Tjian 1986).

In contrast to these very widespread sequence motifs, another sequence element in the *hsp70* gene is shared only with other genes whose transcription is increased in response to elevated temperature. This sequence is found 62 bases upstream of the start site for transcription of the *Drosophila hsp70* gene and at a similar position in other heat-inducible genes (Davidson *et al.* 1983). This heat-shock consensus element is therefore believed to play a critical role in

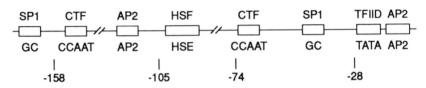

Figure 6.3 Transcriptional control elements in the human *hsp70* gene promoter. The protein binding to a particular site is indicated above the line and the corresponding DNA element below the line. These elements are described more fully in Table 6.2.

144

mediating the observed heat inducibility of transcription of these genes.

In order to confirm that this is the case, it is necessary to transfer this sequence from the *hsp70* gene to another gene, which is not normally heat inducible, and show that the recipient gene now becomes inducible. This was achieved by Pelham (1982) who transferred the heat-shock consensus element onto the non-heat-inducible thymidine kinase (*tk*) gene taken from the eukaryotic virus, herpes simplex. When the hybrid gene was introduced into cells and the temperature subsequently raised, increased thymidine kinase production was detected, showing that the heat-shock consensus element had rendered the *tk* gene inducible by elevated temperature (Fig. 6.4).

This experiment therefore proves that the common sequence element

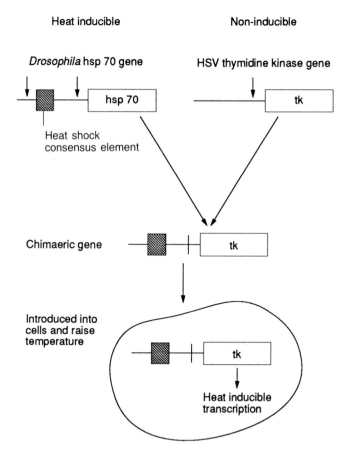

Figure 6.4 Demonstration that the heat-shock consensus element mediates heat inducibility. Transfer of this sequence to a gene (thymidine kinase) which is not normally inducible renders this gene heat inducible.

found in the heat-inducible genes is responsible directly for their heat inducibility. The manner in which these experiments were carried out also permits a further conclusion with regard to the way in which this sequence acts. Thus, the heat-shock sequence element used by Pelham was taken from the *Drosophila hsp70* gene and, in this cold-blooded organism, would be activated normally by the thermally stressful temperature of 37°C. The cells into which the hybrid gene was introduced, however, were mammalian cells which grow normally at 37°C and only express the heat-shock genes at the higher temperature of 42°C. In these experiments the hybrid gene was induced only at 42°C, the heat-shock temperature characteristic of the cell into which it was introduced, and not at 37°C, the temperature characteristic of the species from which the DNA sequence came. This means that the heat-shock consensus sequence does not possess some form of inherent temperature sensor or thermostat which is set to go off at a particular temperature, since in this case the *Drosophila* sequence would activate transcription at 37°C, even in mammalian cells. Rather, it must act by being recognized by a cellular protein which is activated in response to elevated temperature and, by binding to the heat-shock element, produces increased transcription. Evidently, although the elements of this response are conserved sufficiently to allow the mammalian protein to recognize the *Drosophila* sequence, the mammalian protein will, of course, only be activated at the mammalian heat-shock temperature and hence induction will only occur at 42°C.

Hence these experiments not only provide evidence for the importance of the heat-shock consensus element in causing heat-inducible transcription, but also indicate that it acts by binding a protein. Direct evidence that this is the case was provided by Wu (1984), using a technique that involved digestion of the DNA with the

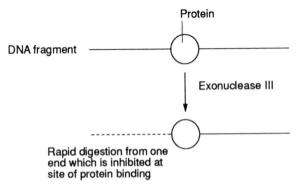

Figure 6.5 Detection of a protein binding to a DNA sequence by inhibition of DNA digestion with exonuclease III.

a) Uninduced cells

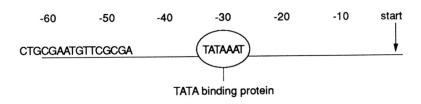

TATA binding protein

b) Induced cells

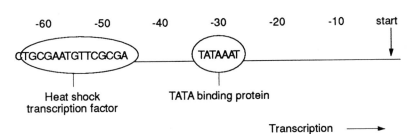

Figure 6.6 Proteins binding to the promoter of the *hsp70* gene before (a) and after (b) heat shock.

enzyme exonuclease III. Although this enzyme will progressively digest DNA, starting at one end, its progress is impeded by the presence of a protein on the DNA, and hence the positions of such proteins can be mapped (Fig. 6.5). When such an analysis is carried out on the upstream regions of the heat-shock genes, a single protein bound to the TATA box region is observed in non-heat-shocked cells (Fig. 6.6a). The binding of this protein is believed to displace a nucleosome and create the DNaseI hypersensitive sites observed in these genes in non-heat-shocked cells (Ch. 5). Hence the potentially activatable state illustrated in Figure 5.24a is created by the binding of this protein.

By contrast, in heat-shocked cells, where transcription of the gene is occurring, an additional protein, which is bound to the heat-shock consensus element, is detectable on the upstream region (Fig. 6.6b). Hence the induction of the heat-shock genes is indeed accompanied by the binding of a protein, known as the heat-shock transcription factor, to the heat-shock consensus sequence, as suggested by the experiments of Pelham (1982). The binding of this factor to a gene whose chromatin structure has already been altered to render it potentially

147

activatable, results in its transcription exactly as discussed in Chapter 5 and illustrated in Figure 5.24. In agreement with this, the purified heat-shock transcription factor can bind to the heat-shock consensus element and stimulate the transcription of the *hsp70* gene in a cell-free nuclear extract while having no effect on the transcription of the non-heat-inducible actin gene (Parker & Topol 1984).

The activation of these heat-inducible genes can therefore be fitted very readily into the Britten and Davidson scheme for the regulation of gene expression. The heat-shock consensus sequence represents the common sequence present in the similarly regulated genes, whereas the heat-shock transcription factor would represent the product of the integrator gene which regulates their expression. Unlike the original model, however, it is clear that the heat-shock transcription factor is not synthesized *de novo* in response to thermal stress, rather it is present in unstressed cells (Parker & Topol 1984) and can activate the heat-shock genes following exposure to elevated temperature even in the presence of protein synthesis inhibitors preventing its *de novo* synthesis (Zimarino & Wu 1987). It is clear, therefore, that upon heat shock a previously inactive transcription factor is activated to a DNA-binding form by a post-translational modification involving an alteration of the protein itself. This activation has been produced recently in an isolated cell-free nuclear extract by elevated temperature, and appears to involve both a temperature-dependent change in the structure of the protein and its modification by phosphorylation (Larson *et al.* 1988).

6.2.3 Other response elements

Although some modification of the Britten and Davidson model, with regard to the activation of the transcription factor itself, is therefore necessary, the heat-shock system does provide a clear example of the role of short common sequences within the promoters of particular genes in mediating their common response to a specific stimulus. A number of similar elements, which are found in the promoters of genes activated by other signals, have now been identified and have been shown to be capable of transferring the specific response to another marker gene (reviewed by Davidson *et al.* 1983). A selection of such sequences is listed in Table 6.3.

As indicated in Table 6.3, these sequences act by binding specific proteins which are synthesized or activated in response to the inducing signal. Such transcription factors are discussed further in Chapter 7. It is noteworthy, however, that many of the sequences in Table 6.3 exhibit dyad symmetry, a similar sequence being found in the 5' to 3'

Table 6.3

Sequences which confer response to a particular stimulus

Consensus sequence	Response to	Protein factor[1]	Genes containing sequences
CTNGAATNTTCTAGA	Heat	Heat shock transcription factor	hsp70, hsp83, hsp27 etc
T/G T/A CGTCA	Cyclic AMP	CREB / ATF	Somatostatin, fibronectin, alpha-gonadotrophin, c-*fos*, hsp70
TGAGTCAG	Phorbol esters	AP1	Metallothionein II A alpha-1-anti-trypsin, collagenase,
GATGTCCATATT AGGACATC	Growth factors in serum	Serum response factor	c-*fos*, *Xenopus* gamma actin
GGTACANNNTGTTCT	Glucocorticoid, progesterone	GR and PR receptors	Metallothionein II A, tryptophan oxygenase, uteroglobin, lysozyme
AGGTCANNNTGACCT	Oestrogen	Oestrogen receptor	Ovalbumin, conalbumin vitellogenin
TCAGGTCATGA CCTGA	Thyroid hormone, retinoic acid	TH and RA receptors	Growth hormone, myosin heavy chain
TGCGCCCGCC	Heavy metals	Not known	Metallothionein genes
AAGTGA	Viral infection	Not known	Interferon alpha and beta, tumour necrosis factor

[1]These protein factors are discussed in Chapter 7

direction on each strand. The oestrogen response element for example, has the sequence:

5' A G G T C A N N N T G A C C T 3'

3' T C C A G T N N N A C T G G A 5'

the two halves of the ten-base palindrome being separated by three random bases. Such symmetry in the binding sites for these transcription factors suggests that they may bind to the site in a dimeric form consisting of two protein molecules.

Sequences that confer response to several different signals have been identified. Exactly as suggested by Britten and Davidson, one gene can possess more than one such element, allowing multiple patterns of regulation. Thus comparison of the sequences listed in Table 6.3 with those contained in the *hsp70* gene, listed in Table 6.2, reveals that, in addition to the heat-shock consensus element, this gene also contains the cyclic AMP response element (CRE) which mediates the induction of a number of genes, such as that encoding somatostatin, in response to treatment with cyclic AMP. Similarly, while genes may share particular elements, flexibility is provided by the presence of other elements in one gene and not another, allowing the induction of a particular gene in response to a given stimulus which has no effect on another gene. For example, although the *hsp70* gene and the metallothionein IIA gene share a binding site for the transcription factor AP2, only the metallothionein gene has a binding site for the glucocorticoid receptor, which confers responsivity to glucocorticoid hormone induction, and hence only this gene is inducible by hormone treatment.

In some cases, the sequence elements that confer response to a particular stimulus can be shown to be related to one another. Thus the sequences mediating response to glucocorticoid or progesterone treatment are similar to that which mediates response to another steroid, namely oestrogen. Similarly, both the oestrogen- and thyroid-hormone-responsive elements contain identical sequences showing dyad symmetry. In the oestrogen-responsive element, however, the two halves of this dyad symmetry are separated by three bases which vary between different genes, whereas in the thyroid-hormone-responsive element the two halves are contiguous (Table 6.4; see Beato 1989 for review). Such similarities are paralleled by a similarity in the individual cytoplasmic steroid receptor proteins which form a complex

Table 6.4

Relationship of consensus sequences conferring responsivity to various hormones

Glucocorticoid/progesterone	GGTACANNNTGTTCT
Oestrogen	AGGTCANNNTGACCT
Thyroid hormone/retinoic acid	TCAGGTCA-------TGACCTGA

N indicates that any base can be present at this position; a dash indicates that no base is present the gap having been introduced to align the sequence with the other sequences.

with each of these hormones and then bind to the corresponding DNA sequence. All of these receptors can be shown to be members of a large family of related DNA-binding proteins whose hormone and DNA binding specificities differ from one another. The exchange of particular regions of these proteins with those of other family members has provided considerable information on the manner in which sequence-specific binding to DNA occurs, and this is discussed in Chapter 7 (Section 7.2.3).

Although the sequence elements shown in Table 6.3 are all involved in the response to particular inducers of gene expression, it seems likely that other short sequence elements or combinations of elements will also be involved in controlling the tissue-specific patterns of expression exhibited by eukaryotic genes. Thus the octamer motif (ATGCAAAT), which is found in both the immunoglobulin heavy- and light-chain promoters, can confer B-cell-specific expression when linked to a non-regulated promoter (Wirth *et al.* 1987). Similarly, short DNA sequences that bind liver-specific transcription factors have been identified in the region of the rat albumin promoter, known to be involved in mediating the liver-specific expression of this gene, and at least one of these sequence elements is also found in the liver-specific α-2 globulin gene (Lichtsteiner *et al.* 1987).

6.2.4   Mechanism of action of promoter regulatory elements

It is clear, therefore, that short DNA sequence elements located near the start site of transcription play an important role in regulating gene expression in eukaryotes. As indicated in Table 6.3 and discussed above, such sequences mediate transcriptional activation by binding a specific protein. This binding may give rise to gene activity in one of two ways (Fig. 6.7). First, as discussed in Chapter 5 and illustrated by the glucocorticoid receptor, binding of a specific protein may result in displacement of a nucleosome and generation of a DNaseI hyper-sensitive site, allowing easy access to the gene for other transcription factors. The direct activation of transcription by such factors constitutes the second mechanism of gene induction, and is illustrated both by the binding of other non-regulated factors to glucocorticoid-regulated genes following binding of the receptor and by the binding of the heat-shock transcription factor to its consensus sequence in the heat-inducible genes. These factors are likely to act by interacting directly with proteins necessary for transcription, such as the TATA box binding factor or RNA polymerase itself. This interaction facilitates the formation of a stable transcription complex, which may enhance the binding of RNA polymerase to the DNA or alter its structure in a manner which increases its activity (for further discussion see Ch. 7,

a) Alteration of chromatin structure

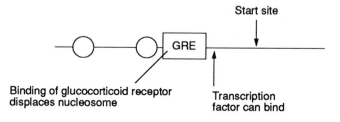

Start site

GRE

Binding site for transcription factor
masked by nucleosome

Start site

GRE

Binding of glucocorticoid receptor
displaces nucleosome

Transcription
factor can bind

b) Interaction with other proteins

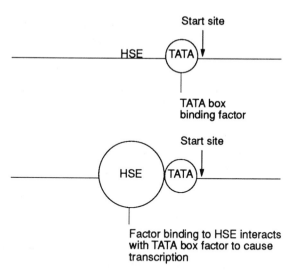

Start site

HSE (TATA)

TATA box
binding factor

Start site

HSE (TATA)

Factor binding to HSE interacts
with TATA box factor to cause
transcription

Figure 6.7 Roles of short sequence elements in gene activation. These elements can either bind a factor that displaces a nucleosome and unmasks a binding site for another factor (a) or can bind a factor that directly activates transcription (b).

Section 7.3.3; Brown 1984, Dynan & Tjian 1985).

It should be noted that these two mechanisms of action are not exclusive. Thus the glucocorticoid receptor, whose binding to a specific DNA sequence displaces a nucleosome, also contains an activation domain capable of interacting with other bound transcription factors (see Ch. 7, Section 7.3.2). Hence, following binding of the receptor, transcription is increased by inter-factor interactions between the receptor and other bound factors.

Short DNA sequence elements act in a similar manner to gene regulatory elements in prokaryotes, and are located at a similar position close to the start site of transcription. Unlike the situation in prokaryotes, however, important elements involved in the regulation of eukaryotic gene expression are also found at very large distances from the site of initiation of transcription. The nature of such regulatory elements will now be discussed.

6.3   ENHANCERS

6.3.1   Regulatory sequences that act at a distance

The first indication that sequences located at a distance from the start site of transcription might influence gene expression in eukaryotes came with the demonstration that sequences over 100 bases upstream of the transcriptional start site of the histone H2A gene were essential for its high-level transcription (Grosschedl & Birnsteil 1980). Moreover, although this sequence was unable to act as a promoter and direct transcription, it could increase initiation from an adjacent promoter element up to one hundredfold when located in either orientation relative to the start site of transcription. Subsequently, a vast range of similar elements have been described in both cellular genes and those of eukaryotic viruses, and have been called enhancers because although they lack promoter activity and are unable to direct transcription themselves, they can dramatically enhance the activity of promoters (for reviews see Serfling *et al.* 1985, Hatzopoulos *et al.* 1988). Hence, if an enhancer element is linked to a promoter, such as that derived from the β-globin gene, the activity of the promoter can be increased several hundredfold. Variation in the position and orientation at which the enhancer element was placed relative to the promoter has led to three conclusions with regard to the action of enhancers. These are:

(a)   An enhancer element can activate a promoter when placed up to several thousand bases from the promoter.

a) Distance

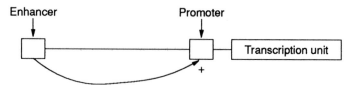

b) Orientation

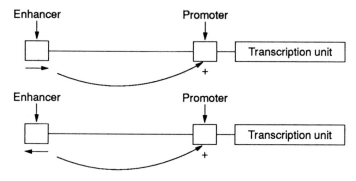

c) Position

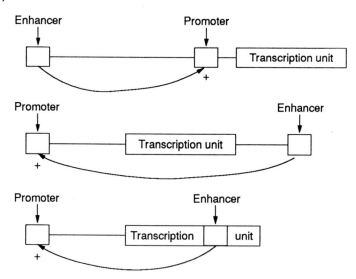

Figure 6.8 Characteristics of an enhancer element which can activate a promoter at a distance (a); in either orientation relative to the promoter (b), and when positioned upstream, downstream, or within a transcription unit (c).

(b) An enhancer can activate a promoter when placed in either orientation relative to the promoter.

(c) An enhancer can activate a promoter when placed upstream or downstream of the transcribed region, or within an intervening sequence which is removed from the RNA by splicing (see Ch. 4, Section 4.2).

These characteristics constitute the definition of an enhancer and are summarized in Figure 6.8.

6.3.2 Tissue-specific activity of enhancers

Following the discovery of enhancers, it was rapidly shown that many genes expressed in specific tissues also contained enhancers. Such enhancers frequently exhibited a tissue-specific activity, being able to enhance the activity of other promoters only in the tissue in which the gene from which they were derived is normally active and not in other tissues (Fig. 6.9). The tissue-specific activity of such enhancers acting on their normal promoters is likely, therefore, to play a critical role in mediating the observed pattern of gene regulation.

Thus, as previously discussed (Ch. 2, Section 2.4), the genes encoding the heavy and light chains of the antibody molecule contain an enhancer located within the large intervening region separating the regions encoding the joining and constant regions of these molecules. When this element is linked to another promoter, such as that of the β-globin gene, it increases its activity dramatically when the hybrid gene is introduced into B-cells. In contrast, however, no effect of the enhancer on promoter activity is observed in other cell types, such as fibroblasts, indicating that the activity of the enhancer is tissue specific (Gillis et al. 1983).

Similar tissue-specific enhancers have also been detected in genes expressed specifically in the liver (α-foetoprotein, albumin, α-1-antitrypsin), the endocrine and exocrine cells of the pancreas (insulin, elastase, amylase), the pituitary gland (prolactin, growth hormone), and many other tissues. The tissue-specific activity of these enhancer elements is likely to play a crucial role in the observed tissue-specific pattern of expression of the corresponding gene. Thus in the case of the insulin gene, early experiments involving linkage of different upstream regions of the gene to a marker gene, and subsequent introduction into different cell types, identified a region approximately 250 bases upstream of the transcriptional start site as being of crucial importance in producing high-level expression in pancreatic endocrine cells (Walker et al. 1983). This position corresponds exactly to the position of the tissue-specific enhancer (Edlund et al. 1985), indicating

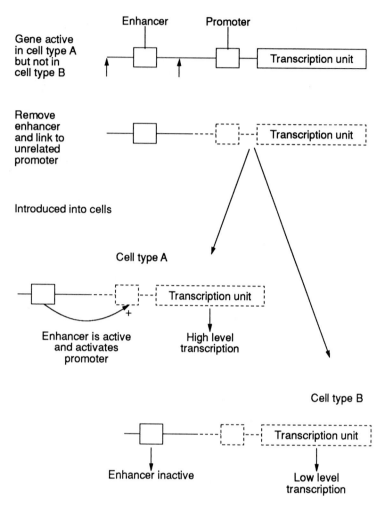

Figure 6.9 A tissue-specific enhancer can activate the promoter of its own or another gene only in one particular tissue and not in others.

the importance of this element in gene regulation. Similarly, mutation of conserved sequences within the tissue-specific enhancers of genes expressed in the exocrine cells of the pancreas, such as elastase and chymotrypsin, abolishes the tissue-specific pattern of expression of these genes (Boulet *et al.* 1986).

In the case of the insulin gene, the importance of the enhancer element in producing tissue-specific gene expression was further demonstrated by experiments in which this enhancer (together with its adjacent promoter) was linked to the gene encoding the large T antigen of the eukaryotic virus SV40, whose production can be measured

ENHANCERS

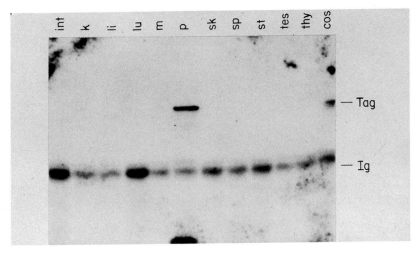

Figure 6.10 Assay for expression of a hybrid gene in which the SV40 T-antigen protein-coding sequence is linked to the insulin gene enhancer and promoter. The gene was introduced into a fertilized egg and a transgenic mouse containing the gene in every cell of its body isolated. Expression of the T antigen is assayed by immunoprecipitation of protein from each tissue with an antibody specific for T antigen. Note that expression of the T antigen is detectable only in the pancreas (p) and not in other tissues, indicating the tissue-specific activity of the insulin gene enhancer. The track labelled 'cos' contains protein isolated from a control cell line expressing T antigen. The Ig band in all tracks is derived from the immunoglobulin antibody used to precipitate the T antigen.

readily using a specific antibody. The resulting construction was introduced into a fertilized mouse egg and the expression of large T antigen analysed in all tissues of the transgenic mouse that developed following the return of the egg to the oviduct. Expression of large T was detectable only in the pancreas and not in any other tissue (Fig. 6.10) and was observed specifically in the β cells of the pancreatic islets which produce insulin (Fig. 6.11; Hanahan 1985). The enhancer of the insulin gene is therefore capable of conferring the specific pattern of insulin gene expression on an unrelated gene *in vivo*.

Hence, enhancer elements constitute another type of DNA sequence which is involved in the activation of genes in a particular tissue or in response to a particular stimulus, as envisaged by the Britten and Davidson model. In agreement with this idea, the binding of cell-type specific proteins to the enhancer element has been demonstrated for many different enhancers, including those in the immunoglobulin (Sen & Baltimore 1986) and insulin (Ohlsson & Edlund 1986) genes.

Therefore, in many cases the tissue-specific expression of a gene will be determined both by the enhancer element and sequences adjacent to the promoter. In the liver-specific pre-albumin gene, for example, gene activity is controlled both by the promoter itself, which is active

Insulin   T antigen

Glucagon   Somatostatin

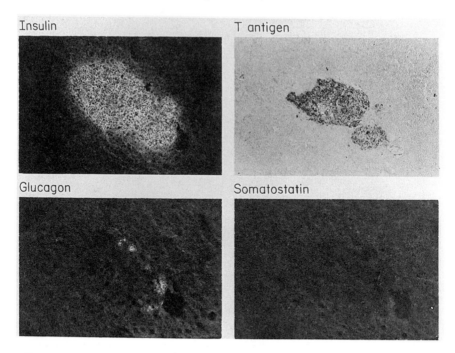

Figure 6.11 Immunofluorescence assay of pancreas preparations from the transgenic mice described in the legend to Figure 6.10 with antibodies to the indicated proteins. Note that the distribution of T antigen parallels that of insulin and not that of the other pancreatic proteins.

only in liver cells, and by an upstream enhancer element, which activates any promoter approximately tenfold in liver cells and not at all in other cell types (Costa *et al.* 1986). Similarly, in the immuno-globulin genes, when the enhancer and the promoter itself are separated, both exhibit B-cell-specific activity in isolation but the maximal expression of the gene is observed only when the two elements are brought together (Garcia *et al.* 1986).

The importance of enhancer elements in the regulation of gene expression therefore necessitates consideration of the mechanism by which they act.

6.3.3 *Mechanism of action of enhancers*

In considering the nature of enhancers, we have drawn distinction between these elements, which act at a distance, and the sequences discussed in Section 6.2, which are located immediately adjacent to the start site of transcription. In fact, however, closer inspection of the sequences within enhancers indicates that they are often composed of the same sequences found adjacent to promoters. For example, the

Figure 6.12 Protein-binding sites in the immunoglobulin heavy-chain gene enhancer. O indicates the octamer motif discussed in the text.

immunoglobulin heavy-chain enhancers contain the octamer motif (ATGCAAAT) which, as discussed previously, is also found in the immunoglobulin promoters. The promoter and enhancer elements bind the identical B-cell-specific transcription factor (as well as a related protein found in all cell types) and play an important role in the B-cell-specific expression of the gene. Interestingly, within the enhancer the octamer motif is found within a modular structure containing binding sites for several different transcription factors, which act together to activate gene expression (Sen & Baltimore 1986; Fig. 6.12). In turn these modules can be subdivided into short DNA sequences, termed enhansons, which are the fundamental units of enhancer function (for a review see Dynan 1989).

The close relationship of enhancer and promoter elements is further illustrated by the *Xenopus hsp70* gene, in which multiple copies of the heat-shock consensus element are located at positions far upstream of the start site and function as a heat-inducible enhancer element when transferred to another gene (Bienz & Pelham 1986).

Enhancers therefore appear to consist of sequence motifs, which are also present in similarly regulated promoters and may be present within the enhancer associated with other control elements or in multiple copies. Indeed, the heat-shock consensus motif of the *Drosophila hsp70* gene, which we have used as the basic example of a promoter motif, has been shown to function as an enhancer when multiple copies are placed at a position well upstream of the transcriptional start site (Bienz & Pelham 1986).

It seems likely, therefore, that enhancers may activate gene expression by either or both of the mechanisms described previously for promoter elements, namely a change in chromatin structure leading to nucleosome displacement or by direct interaction with the proteins of the transcriptional apparatus. In the case of chromatin structure changes, it is readily apparent that such changes caused by a protein binding to the enhancer could be propagated over large distances in both directions, causing the observed distance, position, and orientation independence of the enhancer. In agreement with this possibility, DNaseI hypersensitive sites have been mapped within a number of enhancer elements, including the immunoglobulin enhancer, and the nucleosome-free gap in the DNA of the eukaryotic virus SV40 (see Fig. 5.23) is located at the position of the enhancer.

At first sight, models involving the binding of protein factors to the enhancer followed by direct interaction with proteins of the transcriptional apparatus are more difficult to reconcile with the action at a distance characteristic of enhancers. None the less, the binding to enhancers of very many proteins crucial for transcriptional activation (reviewed by Hatzopoulos *et al.* 1988) suggests that enhancers can indeed function in this manner. Models to explain this postulate that the enhancer serves as a site of entry for a regulatory factor (see Dynan & Tjian 1985). The factor would then make contact with the promoter-

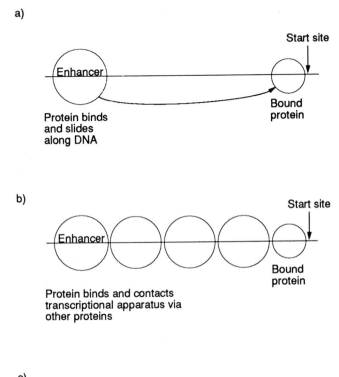

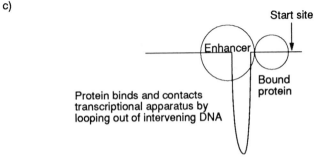

Figure 6.13 Possible models for the action of enhancers located at a distance from the activated promoter.

bound transcriptional apparatus either by sliding along the DNA, or via a continuous scaffold of other proteins, or by the looping out of the intervening DNA (Fig. 6.13).

Of these possibilities, both the sliding model and the continuous scaffold model are difficult to reconcile with the observed large distances over which enhancers act. Similarly, these models cannot explain the observations of Atchison & Perry (1986) who found that the immunoglobulin enhancer activates equally two promoters placed 1.7 kb and 7.7 kb away on the same DNA molecule, since they would postulate that sliding or scaffolded molecules would stop at the first promoter (Fig. 6.14). Moreover, it has been shown recently that an enhancer can act on a promoter when the two are located on two separate DNA molecules linked only by a protein bridge, which would disrupt sliding or scaffolded molecules (Muller *et al.* 1989).

Such observations are explicable, however, via a model in which proteins bound at the promoter and enhancer proteins have an affinity for one another and make contact via looping out of the intervening DNA. Moreover, such a model can explain readily the critical importance of DNA structure on the action of enhancers. Thus it has been shown that removal of precise multiples of ten bases (one helical turn) from the region between the SV40 enhancer and its promoter has no effect on its activity, but deletion of DNA corresponding to half a helical turn disrupts enhancer function severely (Takahasi *et al.* 1986).

6.3.4 Positive and negative action of enhancer elements

Thus far we have assumed that enhancers act in an entirely positive manner. In a tissue containing an active enhancer-binding protein, the enhancer will activate a promoter, whereas in other tissues where the protein is absent or inactive, the enhancer will have no effect (Fig.

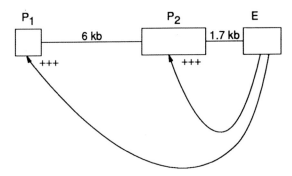

Figure 6.14 An enhancer activates an adjacent promoter and a more distant one equally well.

161

a) Positive action

Inactive tissue or cell

◯ Factor absent or inactive

```
      ┌─┐                    ┌─┐
──────┤ ├────────────────────┤ ├
      └─┘                    └─┘
       E                      P
```

Active tissue or cell

Active factor
binds to
enhancer

```
      ┌─┐                    
      │+│                    ++
──────┤ ├────────────────────►┌─┐
      └─┘                    └─┘
       E                      P
```

b) Negative action

Inactive tissue or cell

Negative factor binds
to enhancer and
prevents binding of
active positive factor

```
                ┌─┐
                │+│
                └─┘
      ┌─┐                    
      │-│                    
──────┤ ├────────────────────┌─┐
      └─┘                    └─┘
       E                      P
```

Active tissue or cell

⊖ Negative factor absent or inactive

Positive factor binds

```
      ┌─┐                    
      │+│                    ++
──────┤ ├────────────────────►┌─┐
      └─┘                    └─┘
       E                      P
```

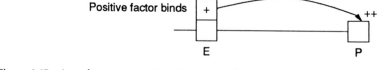

Figure 6.15 An enhancer can act in a tissue-specific manner either by binding a positive factor, which is present in an active form only in a few tissues (a), or by binding a negative factor, which is inactivated in tissues where the enhancer is active (b).

6.15a). Such a mechanism does indeed appear to operate for the majority of enhancers which, when linked to promoters, activate gene expression in one or a few cell types and have no effect in other cell types. In contrast, however, some enhancers produce specific patterns of gene expression by relieving a negative effect. In these cases, it appears that the enhancer binds a factor present in all tissues which inhibits the promoter. Following induction, or in a particular tissue, this negative factor is inactivated and falls off the enhancer, allowing a positive-acting factor to bind and activate gene expression. This positive-acting factor is present in all cells, but cannot normally bind to the DNA because of the presence of the negative repressor factor (Fig. 6.15b). Evidence has been obtained that the enhancer element in the human β-interferon gene acts in this manner (Goodbourn *et al.* 1986). This enhancer, which mediates the induction of the gene in response to viral infection, contains a negative element which can repress gene expression and which binds two proteins present in all cells. Following viral infection, however, these proteins dissociate and another factor binds to an adjacent region of the enhancer, activating gene expression (Zinn & Maniatis 1986).

A more extreme example of an enhancer-like sequence which acts in an entirely negative manner is found in the cellular oncogene c-*myc* (reviewed by Linzer 1985). This element, which is referred to as a silencer acts in either orientation to inhibit activity of distant promoters without binding a regulatory protein, and hence appears to have an inherently negative effect. In cells expressing the c-*myc* gene, the negative effect of this element is presumably relieved by the binding of

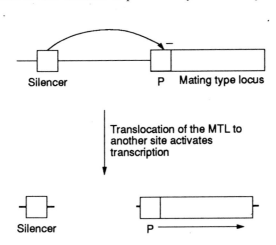

Figure 6.16 The negative effect of the silencer element on transcription of the yeast mating type locus is eliminated when it is separated from the promoter by the DNA rearrangement event.

a regulatory protein, and can also be eliminated by its deletion, which is observed in some cancer cells expressing the c-*myc* gene (see Ch. 8, Section 8.3). Similar silencer elements have also been observed in yeast, located approximately 2 kb from the promoters of the repressed mating type loci (see Ch. 2, Section 2.4). These silencers play a crucial role in organizing this region into the tightly packed structure characteristic of non-transcribed DNA (Nasmyth 1982, Brand *et al.* 1985). In this case, relief from the negative effect of the silencer sequences is provided by the physical separation of the repressed locus from the silencer sequence and its translocation to another site, where it becomes active (Fig. 6.16).

In summary, therefore, whether acting positive or negatively, enhancers play a critical role in the regulation of eukaryotic gene expression, often acting in concert with related regulatory elements located adjacent to the promoter itself.

6.4   ROLE OF REPEATED SEQUENCES

6.4.1   Repeated DNA

In eukaryotes, as in prokaryotes, most structural genes encoding individual proteins are present in only one or a few copies in each genome. In addition, however, the eukaryotic genome contains repeated sequences which, as their name implies, are present in many copies in the genome (for a review see Jelinek & Schmid 1982). Of particular interest, from the point of view of gene regulation, is the class of repeated sequence known as SINES or short interspersed repeats (reviewed by Rogers 1985). Examples of this type of sequence are found in all higher eukaryotes and include the Alu sequence in humans, the B1 and B2 repeats in the mouse, and the suffix element in *Drosophila*. Such sequences are generally about 200–300 bp long and are present in up to 100 000 copies per genome. As their name implies, these copies are not present as tandem repeats but are found interspersed amongst non-repeated DNA (Fig. 6.17).

Often these elements are found in the intervening sequences or the non-translated region of protein-coding genes, and are transcribed as part of the gene by RNA polymerase II (Fig. 6.18a). In addition,

Figure 6.17   Interspersion of SINE repeated sequences (R) with unique DNA.

a) RNA polymerase II

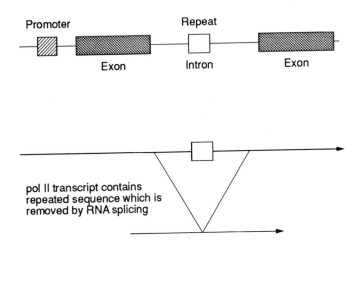

b) RNA polymerase III

pol III transcribes repeat alone ⟶

Figure 6.18 Transcription of SINE elements as part of large RNA polymerase II transcripts (a), or independently as small RNA polymerase III transcripts containing only the repeated element (b).

however, a subset of SINE repeats can be transcribed independently, by RNA polymerase III (Fig. 6.18b).

6.4.2 Role of repeated sequences in gene expression

In the original model of Britten & Davidson (1969) it was suggested that repeated sequences might play the primary role in co-ordinating the expression of unlinked genes, by acting as a common regulatory element adjacent to co-ordinately regulated genes. Thus, since such repeated elements are known to be interspersed amongst non-repeated structural genes, genes regulated in parallel would be adjacent to copies of a particular repeat and their co-ordinate activation would be mediated by the common repeated sequence. As we have seen, however, (Sections 6.2 & 6.3) considerable evidence has been obtained that this role is in fact fulfilled by short elements, less than 50 bases in length, located adjacent to promoters or within enhancers. In contrast, no evidence is available suggesting that SINE elements can act in this manner, and it is unlikely that they do so.

In the course of testing this hypothesis, however, much information has been accumulated, and some of it suggests that these repeated sequences may play other roles in gene regulation. It has been established that, in many cases, a number of different transcription units containing a particular repeated sequence are transcribed in a particular tissue. Such examples include the transcription of multiple RNAs containing the B2 repeats in early mouse development (Murphy *et al*. 1983; Fig. 6.19), and small transcripts containing the so-called identifier, or ID, repeat are detectable specifically in rat brain (Sutcliffe *et al*. 1984).

In many cases, the repeated transcripts that are expressed in a tissue-specific manner are the small RNAs produced by RNA polymerase III transcription of individual repeated elements, rather than the larger repeat-containing transcripts produced by polymerase II. Thus, contrary to earlier suggestions, it now appears that only polymerase III transcripts of the ID repeat are confined to rat brain, larger polymerase II transcripts being found in the nuclear RNA of all tissues (Owens *et al*. 1985, see Fig. 6.20).

Although the tissue-specific transcription of these repeats by polymerase III represents an aspect of gene regulation that needs to be understood (see Section 6.5), it has also been suggested that such transcription might play a role in regulating the activity of genes

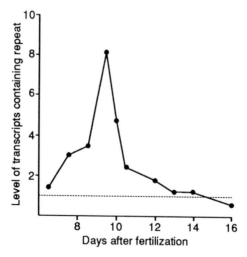

Figure 6.19 Expression of transcripts containing the B2 repeat in mouse development.

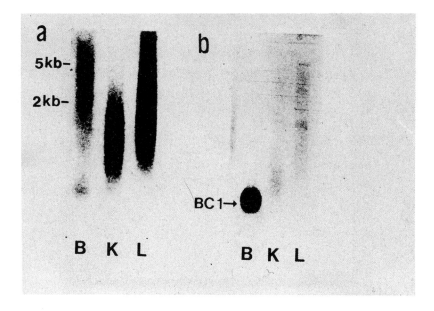

Figure 6.20 Northern blot showing (a) that large transcripts containing the ID repeat are detectable in the nuclear RNA of brain (B), kidney (K), and liver (L); while (b) the small ID transcript BC1 is present only in the cytoplasmic RNA of the brain.

transcribed by polymerase II. Two suggestions have been put forward as to how this might occur.

CHROMATIN STRUCTURE REGULATION

It has been observed that in some cases the tissue-specific transcription of a repeated element by polymerase III occurs in the same tissues that transcribe an adjacent polymerase II gene. For example, an Alu repeat adjacent to the human ε-globin gene is transcribed only in erythroid tissues transcribing the gene (Allan & Paul 1984; Fig. 6.21). Thus it has been suggested that the repeated element could be recognized by a tissue-specific transcription factor and RNA polymerase III even within the tightly packed chromatin structure characteristic of inactive genes

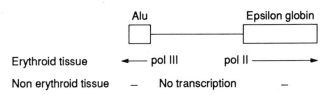

Figure 6.21 Transcription of the ε-globin gene and its adjacent Alu repeated sequence is observed only in erythroid tissue.

Inactive tissue

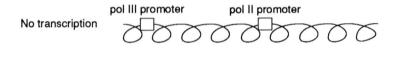

Active tissue

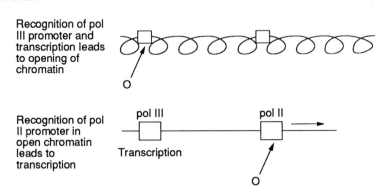

Figure 6.22 Possible role of polymerase III transcription units in the parallel regulation of adjacent polymerase II transcription units. A tissue-specific factor is postulated to recognize the polymerase III promoter within the repeated sequence, even in tightly packed chromatin. Transcription of this repeat by polymerase III opens up the adjacent chromatin, allowing transcription factors access to the polymerase II promoter, and transcription of the polymerase II gene begins.

(Fig. 6.22). Transcription of this repeat would then open up the chromatin in the region, as described in Chapter 5, allowing transcription from the adjacent polymerase II promoter. Although this model is appealing, there is no direct evidence in its favour, and it is possible that the observed association is entirely casual or, indeed, that tissue-specific regulatory processes acting on the polymerase II gene activate the adjacent polymerase III transcription unit directly or indirectly.

POST-TRANSCRIPTIONAL CONTROL

As discussed in Chapter 3 (Section 3.2.2) a more recent version of the Britten and Davidson model (Davidson & Britten 1979) suggested that regulation of gene expression could occur at a post-transcriptional level by deciding which RNAs should be processed to mature transcripts (see Fig. 3.7). This would occur via the tissue-specific transcription of short repeated units by polymerase III in a particular tissue, which would allow the processing in that tissue of large polymerase II RNAs

Tissue A

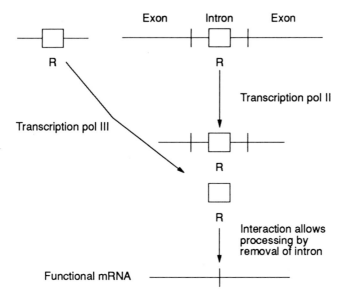

Transcription pol III

Transcription pol II

Functional mRNA

Interaction allows
processing by
removal of intron

Tissue B

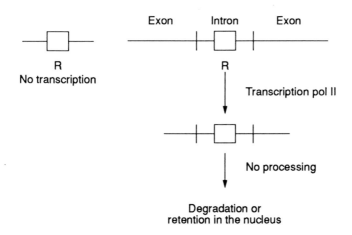

Figure 6.23   Davidson and Britten model for the involvement of repeated sequences in
gene regulation. Repeat-containing genes are postulated to be transcribed in all tissues,
but the transcript can only be processed to a functional mRNA in a tissue (A) where the
isolated repeat is also transcribed by RNA polymerase III. Hence expression of RNA
polymerase II genes is regulated by the selective transcription, in different tissues, of
isolated repeated elements by RNA polymerase III.

containing the identical repeat. These large transcripts would be transcribed in all tissues but would not be processed in tissues lacking the small RNAs, producing a tissue-specific pattern of messenger RNA accumulation (Fig. 6.23).

It is possible that such a mechanism may be involved in some of the cases of post-transcriptional regulation which have been described (see Ch. 4). Thus the ID repeat has been shown recently to confer a growth-dependent pattern of regulation on a transcription unit into which it is inserted, and to do so by affecting post-transcriptional events rather than by increasing transcription (Glaichenhaus & Cuzin 1987). It remains to be seen, however, whether such regulation is achieved by the mechanism postulated by Britten and Davidson.

It is possible, therefore, that repeated elements may play some role in gene regulation, either at the level of chromatin structure or of post-transcriptional control. It is clear, however, that they do not play a role in the co-ordinate induction of unlinked genes, this being effected by the much shorter common sequence elements found adjacent to the promoters of these genes and within enhancers.

6.5 REGULATION OF TRANSCRIPTION BY RNA POLYMERASES I AND III

6.5.1 RNA polymerase I

RNA polymerase I is responsible for the transcription of the tandem arrays of genes encoding ribosomal RNA, such transcription constituting about one-half of total cellular transcription. As with RNA polymerase II, regulatory sequences upstream of the start site of transcription are involved in controlling the transcription of these genes, with sequences from the initiation site to -50 playing an essential role, and sequences farther upstream having a modulatory function (for reviews see Sommerville 1984, La Thangue & Rigby 1988). Although the rate of transcription of the ribosomal genes can be altered by various stimuli, such as viral infection or changes in cellular growth rate, the processes whereby this occurs are not yet understood.

6.5.2 RNA polymerase III

The differences in transcription of repeated sequences by RNA polymerase III, which occur in different tissues (Section 6.4.2), clearly indicate that transcription by this polymerase must be regulated. Of the many types of transcription unit that are transcribed by polymerase

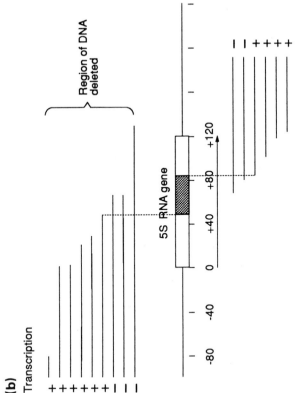

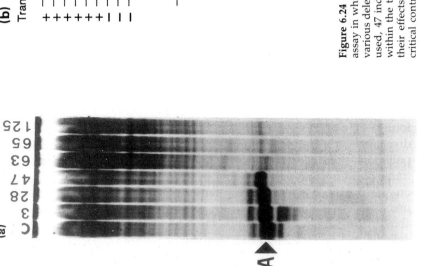

(b)

Transcription

Region of DNA deleted

5S RNA gene

-80   -40   0   +40   +80   +120

Figure 6.24 Effect of deletions in the 5S rRNA gene on its expression. (a) Transcription assay in which the production of 5S RNA (arrowed) by an intact control 5S gene (C) and various deleted 5S genes is assayed. The numbers indicate the end point of each deletion used, 47 indicating that the deletion extends from the upstream region to the 47th base within the transcribed region, etc. (b) Summary of the extent of the deletions used and their effects on transcription. The use of these deletions allows the identification of a critical control element (boxed) within the transcribed region of the 5S gene.

5S gene

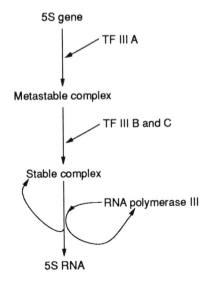

Figure 6.27   Stages in the formation of a stable transcription complex on the 5S gene.

that encoding the 7SL RNA of the signal recognition particle, and the Alu repeated sequences. These internal sequences are recognized by the transcription factors TFIIIB and TFIIIC discussed above. In some genes such internal sequences operate in conjunction with other sequences upstream of the transcribed region which may play the predominant role in individual cases (reviewed by Sollner-Webb 1988).

Hence, genes transcribed by RNA polymerase III differ from those transcribed by RNA polymerases I and II in having sequences within the transcribed region which are involved in the initiation of transcription.

In the 5S rRNA system, the fact that, unlike upstream sequences, such internal elements will be present in both the template DNA and the transcribed RNA (Fig. 6.25) is exploited in a unique regulatory mechanism. The toad, *Xenopus laevis*, contains two types of 5S genes: the oocyte genes, which are transcribed only in the developing oocyte before fertilization, and the somatic genes, which are transcribed in

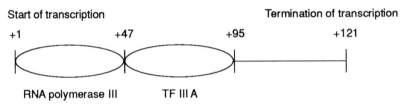

Figure 6.28   Binding of TFIIIA and of RNA polymerase III to the 5S gene.

+46   +98
TCGGAAGCCAAGCAGGGTCGGGCCTGGTTAGTACTTGGATGGGAGACCGCCTGG
        G T

Figure 6.29 Sequence of the internal control element of the *Xenopus* somatic 5S genes. The two base changes in the oocyte 5S genes which result in a lower affinity for TFIIIA are indicated below the somatic sequence.

cells of the embryo and adult. The internal control region of both these types of genes bind TFIIIA, whose binding is necessary for their transcription. However, sequence differences between the two types of gene (Fig. 6.29) result in a higher affinity of TFIIIA factor for the somatic compared to the oocyte genes. Hence, the oocyte genes are only transcribed in the oocyte, where there are abundant levels of TFIIIA, and not in other cells, where the levels are only sufficient for activity of the somatic genes. In the developing oocyte TFIIIA is synthesized at high levels and transcription of the oocyte genes begins. As more and more 5S RNA molecules containing the TFIIIA binding site accumulate in the maturing oocyte, they bind the transcription factor. This factor is thus sequestered with the 5S RNA into storage particles and is unavailable for transcription of the 5S rRNA genes.

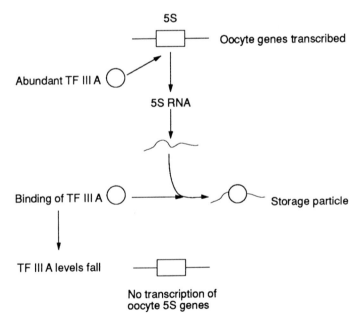

Figure 6.30 Sequestration of the TFIIIA factor by binding to 5S RNA made in the oocyte results in a fall in the level of the factor, switching off transcription of the oocyte-specific 5S genes. The somatic 5S genes, which have a higher affinity binding site for TFIIIA, continue to be transcribed.

175

proteins on the promoter of the mouse albumin gene. *Cell* **51**, 9963–73.

Linzer, D. 1985. Negative transcriptional regulation of c-*myc*. *Trends in Genetics* **2**, 195.

McKnight, S. & R. Tjian 1986. Transcriptional selectivity of viral genes in mammalian cells. *Cell* **46**, 795–805.

Maniatis, T., S. Goodbourn & J. A. Fischer 1987. Regulation of inducible and tissue-specific gene expression. *Science* **236**, 1237–45.

Maniatis, T., E. F. Fritsch, J. Lauer & R. M. Lawn 1980. The molecular genetics of human hemoglobins. *Annual Review of Genetics* **14**, 145–78.

Muller, H.-P., J. W. Soga & W. Schaffner 1989. An enhancer stimulates transcription in *trans* when linked to the promoter via a protein bridge. *Cell* **58**, 767–77.

Murphy, D., P. M. Brickell, D. S. Latchman, K. Willison & P. W. J. Rigby 1983. Transcripts regulated during normal embryonic development and oncogenic transformation share a common repetitive element. *Cell* **35**, 865–71.

Nasmyth, K. A. 1982. Molecular genetics of yeast mating type. *Annual Review of Genetics* **16**, 439–500.

Ohlsson, H. & T. Edlund 1986. Sequence specific interactions of nuclear factors with the insulin gene enhancer. *Cell* **45**, 35–44.

Owens, G. P., N. Chaudhari & W. E. Hahn 1985. Brain identifier sequence is not restricted to brain: similar abundance in nuclear RNA of other organs. *Science* **229**, 1263–5.

Parker, C. S. & J. Topol 1984. A *Drosophila* RNA polymerase II transcription factor binds to the regulatory site of an *hsp70* gene. *Cell* **37**, 273–83.

Pelham, H. R. B. 1982. A regulatory upstream promoter element in the *Drosophila hsp70* heat-shock gene. *Cell* **30**, 517–28.

Rogers, J. H. 1985. The origin and evolution of retroposons. *International Review of Cytology* **93**, 187–279.

Sakonju, S., D. F. Bogenhagen & D. D. Brown 1980. A control region in the center of the 5S gene directs specific initiation of transcription. II. The 3' border of the region. *Cell* **19**, 27–35.

Schmitz, A. & D. Galas 1979. The interaction of RNA polymerase and *lac* repressor with the *lac* control region. *Nucleic Acids Research* **6**, 111–37.

Sen, R. & D. Baltimore 1986. Multiple nuclear factors interact with the immunoglobulin enhancer sequences. *Cell* **46**, 705–16.

Serfling, E., M. Jasin & W. Schaffner 1985. Enhancers and eukaryotic gene transcription. *Trends in Genetics* **1**, 224–30.

Sollner-Webb, B. 1988. Suprises in polymerase III transcription. *Cell* **52**, 153–54.

Sommerville, J. 1984. RNA polymerase I promoters and cellular transcription factors. *Nature* **310**, 189–90.

Sutcliffe, J. G., R. J. Milner, J. M. Gottesfeld & W. Reynolds 1984. Control of neuronal gene expression. *Science* **225**, 1308–15.

Takahashi, K., M. Vigneron, H. Matthes, A. Wildeman, M. Zeake & P. Chambon 1986. Requirement of stereospecific alignments for initiation from the simian virus 40 early promoter. *Nature* **319**, 121–6.

Walker, M. D., T. Edlund, A. M. Boulet & W. J. Rutter 1983. Cell specific expression controlled by the 5' flanking region of the insulin and chymotrypsin genes. *Nature* **306**, 557–61.

Williams, G. T., T. K. McClanahan & R. I. Morimoto 1989. Ela transactivation of the human *hsp70* promoter is mediated through the basal transcriptional complex. *Molecular and Cellular Biology* **9**, 2574–87.

REFERENCES

Wirth, T., L. Staudt & D. Baltimore 1987. An octamer oligonucleotide upstream of a TATA motif is sufficient for lymphoid specific promoter activity. *Nature* **329**, 174–8.

Wu, C. 1984. Two protein-binding sites in chromatin implicated in the activation of heat-shock genes. *Nature* **309**, 229–34.

Zimarino, V. & C. Wu 1987. Induction of sequence-specific binding of *Drosophila* heat-shock activator protein without protein synthesis. *Nature* **327**, 727–30.

Zinn, K. & T. Maniatis 1986. Detection of factors that interact with the human beta-interferon regulatory region *in vivo* by DNAase I footprinting. *Cell* **45**, 611–18.

CHAPTER SEVEN

Transcriptional control – transcription factors

7.1 INTRODUCTION

As discussed in Chapter 6, the expression of specific genes in particular cell types or tissues is regulated by DNA sequence motifs present within promoter or enhancer elements which control the alteration in chromatin structure of the gene that occurs in a particular lineage, or the subsequent induction of gene transcription. It has been assumed for many years that such sequences would act by binding a regulatory protein which was only synthesized in a particular tissue or was present in an active form only in that tissue. In turn, the binding of this protein would result in the observed effect on gene expression.

The isolation and characterization of such factors proved difficult, however, principally because they were present in very small amounts. Hence, even if they could be purified, the amounts obtained were too small to provide much information as to the properties of the protein.

This obstacle was overcome by the cloning of the genes encoding a number of different transcription factors. Two general approaches were used to achieve this. In one approach (Fig. 7.1), exemplified by the work of Kadonga & Tjian (1986), the transcription factor Sp1 was purified by virtue of its ability to bind to its specific DNA binding site. The partial amino-acid sequence of the protein was then obtained from the small amount of material isolated, and was used in conjunction with the genetic code to predict a set of DNA oligonucleotides, one of which would encode this region of the protein. The oligonucleotides were then hybridized to a complementary DNA library prepared from Sp1-containing HeLa cell mRNA. A cDNA clone derived from the Sp1 mRNA must contain the sequence capable of encoding the protein and hence will hybridize to the probe. In this experiment one single clone derived from the Spl mRNA was isolated by screening a library of 1 million recombinants prepared from the whole population of HeLa cell mRNAs (Kadonga *et al.* 1987).

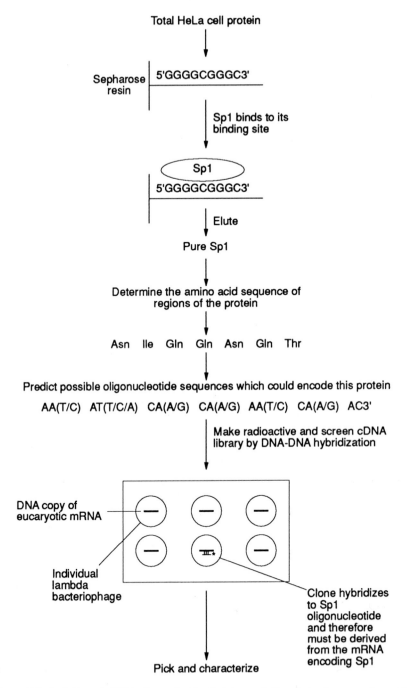

Total HeLa cell protein

Sepharose resin   5'GGGGCGGGC3'

Sp1 binds to its binding site

Sp1

5'GGGGCGGGC3'

Elute

Pure Sp1

Determine the amino acid sequence of regions of the protein

Asn Ile Gln Gln Asn Gln Thr

Predict possible oligonucleotide sequences which could encode this protein

AA(T/C) AT(T/C/A) CA(A/G) CA(A/G) AA(T/C) CA(A/G) AC3'

Make radioactive and screen cDNA library by DNA-DNA hybridization

DNA copy of eucaryotic mRNA

Individual lambda bacteriophage

Clone hybridizes to Sp1 oligonucleotide and therefore must be derived from the mRNA encoding Sp1

Pick and characterize

Figure 7.1 Isolation of cDNA clones for the Sp1 transcription factor by screening with short oligonucleotides predicted from the protein sequence of Sp1. Because several different triplets of bases can code for any given amino acid, multiple oligonucleotides that contain every possible coding sequence are made. Positions at which these oligonucleotides differ from one another are indicated by the brackets containing more than one base.

An alternative, more direct, approach to the cloning of transcription factors is exemplified by the work of Singh *et al.* (1988) on the NF ϰB protein, which is involved in regulating the expression of the immunoglobulin genes in B-cells (Fig. 7.2). As in the previous method, a cDNA library was constructed containing copies of all the mRNAs in a specific cell type. However, the library was constructed in such a way that the sequences within it would be translated into their corresponding proteins. This was achieved by inserting the cDNA into the coding region of the bacteriophage β-galactosidase gene, resulting in the translation of the eukaryotic insert as part of the bacteriophage protein (for a full description of this technique and its other applications see Young & Davis 1983). Most interestingly, these fusion proteins were capable of binding DNA with the same specificity as the original transcription factor encoded by the cloned mRNA. Hence the library could be screened directly with the radiolabelled DNA binding site for a particular transcription factor. A clone containing the mRNA for this factor, and hence expressing it as a fusion protein, bound the labelled DNA and could be identified readily and isolated.

Unlike the previous method, this procedure involves DNA–protein rather than DNA–DNA binding and can be used without prior purification of the transcription factor, provided its binding site is known. Since most factors are identified on the basis of their binding to a particular site, this is not a significant problem and the use of these two methods has resulted in the isolation of the genes encoding a wide variety of transcription factors.

In turn, this has resulted in an explosion of information on these factors (for general reviews see Johnson & McKnight 1989, Mitchell & Tjian 1989). Thus, once the gene for a factor has been cloned, Southern blotting (Section 2.2.4) can be carried out to study the structure of the gene, Northern blotting (Section 1.3.2) can be used to search for RNA transcripts derived from it in different cell types,and related genes expressed in other tissues or other species can be identified.

More importantly, considerable information can be obtained from the cloned gene about the corresponding protein and its activity. Thus, not only can the DNA sequence of the gene be used to predict the amino-acid sequence of the corresponding protein, but the existence of functional domains within the protein with particular activities can also be defined. As described above, if the gene encoding a transcription factor is expressed in bacteria, it continues to bind DNA in a sequence-specific manner. Hence if the gene is broken up into small pieces and each of these is expressed in bacteria (Fig. 7.3), the abilities of each portion to bind to DNA, to other proteins, or to a potential regulatory molecule can be assessed. This mapping can also be achieved by transcribing and translating pieces of the DNA into protein fragments

Make cDNA library in such a way that eukaryotic
mRNA will be expressed in the bacteria

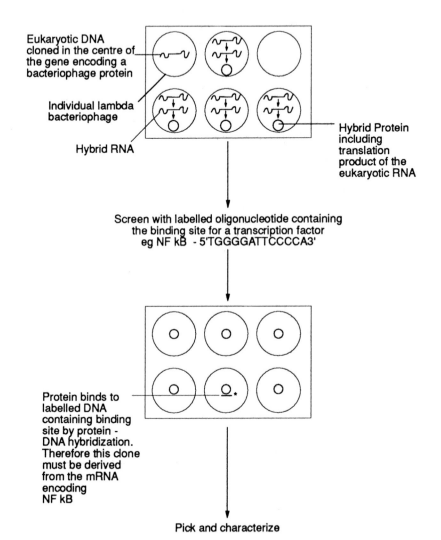

Eukaryotic DNA
cloned in the centre of
the gene encoding a
bacteriophage protein

Individual lambda
bacteriophage

Hybrid RNA

Hybrid Protein
including
translation
product of the
eukaryotic RNA

Screen with labelled oligonucleotide containing
the binding site for a transcription factor
eg NF kB - 5'TGGGGATTCCCCA3'

Protein binds to
labelled DNA
containing binding
site by protein -
DNA hybridization.
Therefore this clone
must be derived
from the mRNA
encoding
NF kB

Pick and characterize

Figure 7.2 Isolation of cDNA clones for the NF κB transcription factor by screening an expression library with a DNA probe containing the binding site for the factor.

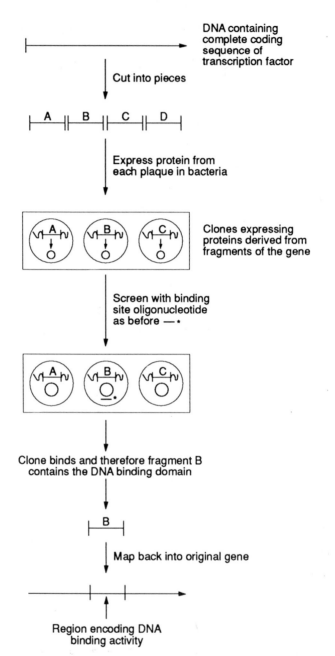

DNA containing complete coding sequence of transcription factor

Cut into pieces

A B C D

Express protein from each plaque in bacteria

A B C

Clones expressing proteins derived from fragments of the gene

Screen with binding site oligonucleotide as before — *

A B C

Clone binds and therefore fragment B contains the DNA binding domain

B

Map back into original gene

Region encoding DNA binding activity

Figure 7.3 Mapping of the DNA-binding region of a transcription factor by testing the ability of different regions to bind to the appropriate DNA sequence when expressed in bacteria.

in the test-tube and testing their activity in the same way.

Each of the domains identified in this way can be altered by mutagenesis of the DNA and subsequent expression of the mutant protein as before. The testing of the effect of these mutations on the activity mediated by the particular domain of the protein will thus allow the identification of the amino acids that are critical for each of the observed properties of the protein.

In this way large amounts of information have accumulated on individual transcription factors. Rather than attempt to consider each factor individually, we will focus on the properties necessary for such a factor and illustrate our discussion by referring to the manner in which these are achieved in individual cases.

It should be evident from the foregoing discussion that the first property such a factor requires is the ability to bind to DNA in a sequence-specific manner, and this is discussed in Section 7.2. Subsequently, the bound factor must influence transcription by interacting with other transcription factors or with the RNA polymerase itself. Section 7.3 therefore considers the means by which DNA-bound transcription factors actually increase transcription. Finally, in the case of transcription factors which activate a particular gene in one tissue only, some means must be found to ensure that the transcription factor is active only in that tissue. Section 7.4 discusses how this is achieved, either by the expression of the gene encoding the factor only in one particular tissue or by a tissue-specific modification which results in the activation of a factor present in all cell types.

7.2 DNA BINDING BY TRANSCRIPTION FACTORS

7.2.1 Introduction

Extensive studies of eukaryotic transcription factors have identified several structural elements, which either bind directly to DNA or which facilitate DNA binding by adjacent regions of the protein (for reviews see Schleif 1988, Struhl 1989). These motifs will be discussed in turn, using transcription factors that contain them to illustrate their properties.

7.2.2 The helix-turn-helix motif

THE HOMEOBOX

The small size and rapid generation time of the fruit fly *Drosophila melanogaster* has led to it being one of the best-characterized organisms

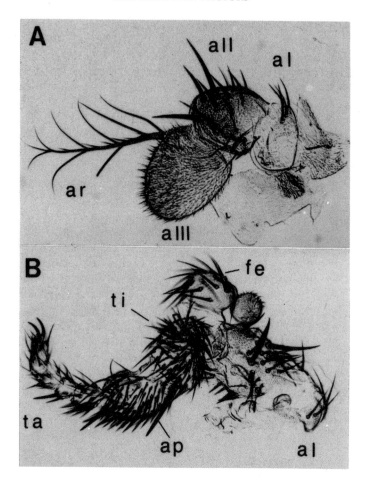

Figure 7.4 Effect of a homeotic mutation, which produces a middle leg (b) in the region that would contain the antenna of a normal fly (a). aI, aII, aIII: 1st, 2nd, and 3rd antennal segments; ar: arista; ta: tarsus; ti: tibia; fe: femur; ap: apical bristle.

genetically, and a number of mutations which affect various properties of the organism have been described. These include a number of mutations which affect the development of the fly, resulting, for example, in the production of additional legs in the position of the antennae (Fig. 7.4). Genes of this type are likely to play a crucial role in the development of the fly and, in particular, in determining the body plan, and are known as homeotic genes (for reviews see Ingham 1988, Scott & Carroll 1987).

The critical role for the products of these genes, identified genetically, suggested that they would encode regulatory proteins which would act at particular times in development to activate or

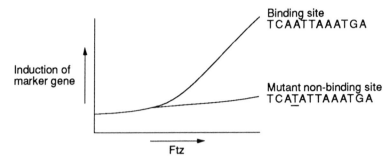

Figure 7.5 Effect of expression of the Ftz protein on the expression of a gene containing its binding site, or a mutated binding site containing a single base pair change which abolishes binding of Ftz.

repress the activity of other genes encoding proteins required for the production of particular structures. This idea was confirmed when the genes encoding these proteins were cloned. Thus, these proteins were shown to be able to bind to DNA in a sequence-specific manner and to be able to induce increased transcription of genes which contained this binding site. Thus in the case of the homeotic gene *fushi tarazu (ftz)*, mutation of which produces a fly with only half the normal number of segments, the protein has shown to bind specifically to the sequence TCAATTAAATGA. When the gene encoding this protein is introduced into *Drosophila* cells with a marker gene containing this sequence, transcription of the marker gene is increased (Jaynes & O'Farrell 1988). This up-regulation is entirely dependent on binding of the Ftz protein to this sequence in the promoter of the marker gene, since a 1 bp change in this sequence, which abolishes binding, also abolishes the induction of transcription (Fig. 7.5).

The product of another homeotic gene, the engrailed protein, binds to the identical sequence to that bound by Ftz. Its binding does not produce increased transcription of the marker gene, however, and indeed it prevents the activation by Ftz. Hence, the expression of Ftz alone in a cell would activate particular genes, whereas Ftz expression in a cell also expressing the engrailed product would have no effect (Fig. 7.6). In this way interacting homeotic gene products expressed in particular cells could control the developmental fate of the cells.

Interestingly, there is evidence that the homeotic genes may be necessary not only for the actual production of a specific cell type but also for the long-term process of commitment to a particular cellular phenotype which was discussed in Chapter 5 (Section 5.2). Thus, in the case of the imaginal discs of *Drosophila* (Section 5.2), commitment to the production of a particular adult structure was maintained through many cell generations in the absence of differentiation. If during this

Antp Arg Lys Arg Gly Arg Gln Thr Tyr Thr Arg Tyr Gln Thr Leu Glu Leu Glu Lys Glu Phe His Phe Asn Arg Tyr Leu Thr Arg Arg Arg
Ubx   Arg   Thr   His
Ftz Ser Thr   Ile

        | Helix | Turn | Recognition helix |

Antp Arg Ile Glu Ile Ala His Ala Leu Cys Leu Thr Glu Arg Gln Ile Lys Ile Trp Phe Gln Asn Arg Arg Met Lys Trp Lys Lys Glu Asn
Ubx   Met Tyr   Glu   Leu   Ile
Ftz   Asp Asn   Ser Ser   Ser   Asp Arg

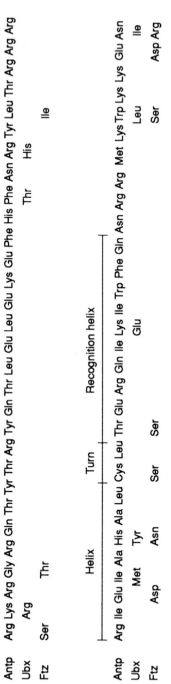

Figure 7.8 Amino-acid sequences of several *Drosophila* homeodomains, showing the conserved helical motifs. Differences between the sequences of the Ubx and Ftz homeodomains from that of Antp are indicated, a blank denotes identity in the sequence. The helix-turn-helix region is indicated.

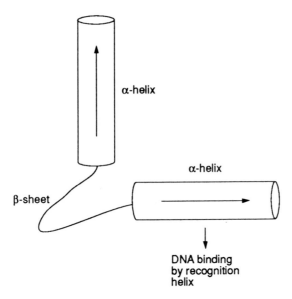

Figure 7.9 The helix-turn-helix motif.

repressor protein, which have been crystallized and whose structures are therefore known with some certainty. By using X-ray crystallography to study the structure of the bacteriophage 434 repressor bound to its binding site, it has been shown that the helix-turn-helix motif does indeed contact DNA. One helix lies across the major groove of the DNA, while the second helix lies partly within the major groove where it can make specific contacts with the bases of the DNA (Fig. 7.10; for further details see Schlief 1988). This second helix (labelled the recognition helix in the homeobox sequence in Fig. 7.8) can thus mediate sequence-specific binding.

The presence of this structure therefore indicates how the homeobox proteins can bind specifically to particular DNA-binding sequences, which is the first step in transcriptional activation of their target genes. Recently, the role of the helix-turn-helix motif in DNA recognition by the homeoproteins has been demonstrated directly (Hanes & Brent 1989) by the finding that a mutation which changes a lysine at position nine of the recognition helix in the Bicoid protein to the glutamine found in the equivalent position of the Antennapedia protein results in the protein binding to DNA with the sequence specificity of an Antennapedia rather than a Bicoid protein. Hence, not only does the helix-turn-helix motif mediate DNA binding, but differences in the precise sequence of this motif in different homeoboxes control the precise DNA sequence to which these proteins bind. Clearly, further immunoglobulins, and mediate the transcriptional activation of these genes (for reviews see Falkner *et al.* 1986, Schaffner 1989). When the genes encoding these proteins were cloned they were found to possess a 150–160 amino-acid sequence that was also found in the mammalian Pit-1 protein, which regulates gene expression in the pituitary by binding to a sequence related to but distinct from the octamer (Ingraham *et al.* 1988), and in the protein encoded by the nematode gene *unc-86*, which is involved in sensory neuron development.

This POU (Pit–Oct–Unc) domain contains both a homeobox-like sequence and a second conserved domain, the POU-specific domain (Fig. 7.11). In Pit-1, as in the *Drosophila* homeobox proteus, the homeodomain alone is sufficient for sequence specific DNA binding, the affinity of binding being enhanced, however, by the POU-specific domain (Theill *et al.* 1989). In contrast, however, Sturm & Herr (1988) showed that, in the case of Oct-1, both parts of the POU domain are required for sequence-specific DNA binding, indicating that the POU homeodomain and the POU-specific domain can form two parts of a DNA-binding element which are held together by a flexible linker sequence.

Unlike the POU-homeodomain, the POU-specific domain cannot form a helix-turn-helix motif, although it can form numerous α-helices. The mechanism by which it binds to the DNA is therefore uncertain and may involve a previously uncharacterized DNA-binding structure.

It is clear, therefore, that the POU proteins represent a new family of proteins, related to the homeobox proteins, which are likely to play a critical role in development, especially in mammals. They may play a particularly critical role in gene regulation in the brain, all the previously characterized POU proteins being expressed in some cells within the brain, while several other novel brain-specific POU proteins have recently been identified (He *et al.* 1989).

7.2.3   The zinc finger motif

THE TWO CYSTEINE–TWO HISTIDINE ZINC FINGER

As discussed in Chapter 6 (Section 6.5.2), one of the earliest gene regulatory systems to be characterized was that of the gene encoding the 5S RNA of the ribosome, in which a transcription factor, TFIIIA, binds to the internal control region of the gene. This transcription factor was amongst the first to be purified (Miller *et al.* 1985). The pure protein was shown to have a periodic repeated structure and to contain between seven and 11 atoms of zinc associated with each molecule of the pure protein.

The basis for this repeated structure was revealed when the gene

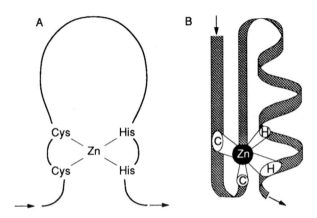

Figure 7.12 Two alternative possible structures of the two cysteine–two histidine zinc finger.

encoding this protein was cloned and used to predict the corresponding amino-acid sequence. This protein sequence contained nine repeats of a 30 amino-acid sequence of the form Tyr/Phe–X–Cys–X–Cys–X$_{2,4}$–Cys–X$_3$–Phe–X$_5$–Leu–X$_2$–His–X$_{3,4}$–His–X$_5$, where X is a variable amino acid. This repeating structure therefore contains two invariant pairs of cysteine and histidine residues which were predicted to bind a single zinc atom, accounting for the multiple zinc atoms bound by the purified protein.

This 30 amino-acid repeating unit is referred to as a zinc finger, on the basis of the proposed structure in which a loop of 12 amino acids, containing the conserved leucine and phenylalanine residues as well as several basic residues, projects from the surface of the protein and is anchored at its base by the conserved cysteine and histidine residues, which directly co-ordinate an atom of zinc (Fig. 7.12A). The binding of zinc by the cysteine and histidine residues has been confirmed directly by X-ray crystallographic analysis of the TFIIIA protein, although an alternative structure for the finger, involving an anti-parallel β-sheet and α-helix between the zinc co-ordination sites, has also been proposed (Fig. 7.12B).

Following the initial identification of the repetitive zinc fingers in TFIIIA, it was hypothesized (Miller *et al.* 1985) that the tips of the fingers would directly contact the 5S DNA, this region being rich in basic amino acids such as arginine and lysine, which could potentially interact with the acidic DNA. The repetitive structure contains multiple fingers which allow the relatively small TFIIIA protein to make repeated contacts with the DNA along the relatively large (approximately 50 bp) 5S DNA internal control region. Subsequent experiments have confirmed this view and have suggested that each TFIIIA finger

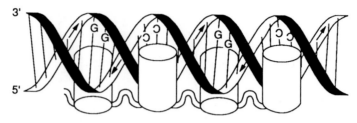

Figure 7.13 Model of the binding of the zinc fingers in TFIIIA to the 5S DNA. Note that adjacent fingers make contact with the DNA from opposite sides of the helix.

binds in the major groove of the DNA helix, interacting with five bases of DNA, or half a helical turn, with successive fingers binding on opposite sides of the helix (Fig. 7.13; reviewed by Klug & Rhodes 1987).

It is clear, therefore, that like the helix-turn-helix motif, the zinc finger represents a protein structure capable of mediating the DNA binding of transcription factors. Although originally identified in the RNA polymerase III transcription factor TFIIIA, this motif has now been identified in a number of RNA polymerase II transcription factors and shown to play a critical role in their ability to bind to DNA and thereby influence transcription (Table 7.1; for reviews see Evans & Hollenberg 1988, Struhl 1989).

Thus three contiguous copies of the 30 amino-acid zinc finger motif are found in the transcription factor Sp1 whose cloning was discussed earlier (Section 7.1). The sequence-specific binding pattern of the intact Sp1 protein can be reproduced by expressing in *E. coli* a truncated protein containing only the zinc finger region, confirming the importance of this region in DNA binding (Kadonga *et al.* 1987). Similarly, the *Drosophila* Kruppel protein, which is vital for proper thoracic and abdominal development, contains four zinc finger motifs. A single mutation, which results in the replacement of the conserved cysteine in one of these fingers by a serine which could not bind zinc, leads to the complete abolition of the function of the protein, resulting in a mutant fly whose appearance is indistinguishable from that produced by complete deletion of the gene (Redemann *et al.* 1988).

As illustrated in Table 7.1, all the zinc finger proteins of the cysteine–histidine type cloned so far have multiple copies of the finger motif, ranging from two in the yeast transcription factor ADR1 to 37 in the *Xenopus* Xfin protein. Interestingly, a single finger from the ADR1 protein is unable to mediate sequence-specific DNA binding, whereas the region of the protein containing both the fingers can do so with the same specificity as the intact protein (Parraga *et al.* 1988). This suggests that interactions between adjacent fingers are essential for sequence-

Table 7.1

Transcriptional regulatory proteins containing Cys$_2$His$_2$ zinc fingers

Organism	Protein	Number of fingers
Drosophila	Kruppel	4
	Hunchback	6
	Snail	4
	Glass	5
Yeast	ADR 1	2
	SW15	3
Xenopus	TFIIIA	9
	Xfin	37
Rat	NGF-1A	3
Mouse	MK1	7
	MK2	9
	Egr 1	3
	Evi 1	10
Human	Sp1	3
	TDF	13

specific binding and provides an explanation for the multi-fingered nature of these proteins.

The zinc finger therefore represents a DNA-binding element which is present in variable numbers in many regulatory proteins. Indeed, the linkage between the presence of this motif and the ability to regulate gene expression is now so strong that, as with the homeobox, it has been used as a probe to isolate the genes encoding new regulatory proteins. The Kruppel zinc finger, for example, has been used in this way to isolate Xfin, a 37 finger protein expressed in the early *Xenopus* embryo (Ruizi-i-Altaba *et al*. 1987). Similarly, the identification of a zinc-finger-containing gene in the sex determining region of the mammalian Y chromosome led to the suggestion that this gene encoded the testis determining factor (TDF), which regulates the expression of genes involved in testis development (Page *et al*. 1987), although further work showed that this was not in fact the case (Burgoyne 1989).

The possible involvement of zinc finger proteins in controlling development in vertebrates, is mirrored in *Drosophila*, where numerous proteins involved in regulating development, such as Kruppel, Hunchback, and Snail, contain zinc fingers. The interactions of these proteins with the homeobox proteins, which contain the alternative

DNA-binding helix-turn-helix motif, is of central importance in the development of *Drosophila* and possibly other organisms.

THE MULTI-CYSTEINE ZINC FINGER

Throughout this work we have noted that the effect of steroid hormones on mammalian gene expression is one of the best-characterized examples of gene regulation. Thus the steroid-regulated genes were amongst the first to be shown to be regulated at the level of gene transcription (Ch. 3) by means of the binding of a specific receptor to a specific DNA sequence (Ch. 6, Section 6.2), resulting in the displacement of a nucleosome and the generation of a DNAseI hypersensitive site (Ch. 5, Section 5.6). When the genes encoding the DNA-binding receptors for the various steroid hormones, such as glucocorticoid and oestrogen, were cloned, they were found to constitute a family of proteins encoded by distinct but related genes. In turn, these proteins were related to other receptors which mediated the response of the cell to hormones such as thyroid hormone or retinoic acid, leading to the idea of an evolutionarily related family of genes encoding hormone receptors, known as the steroid–thyroid hormone

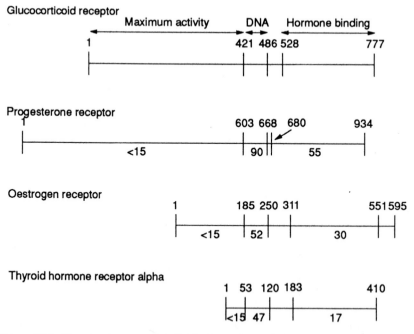

Figure 7.14 Domain structure of individual members of the steroid–thyroid hormone receptor super-family. The proteins are aligned on the DNA-binding domain, which shows the most conservation between different receptors. The percentage homologies in each domain of the receptors to that of the glucocorticoid receptor are indicated.

receptor gene super-family (for reviews see Evans 1988, Beato 1989).

When the detailed structures of the members of this family were compared (Fig. 7.14) it was found that each had a multi-domain structure, which included a central highly conserved domain. On the basis of experiments in which truncated versions of the receptors were introduced into cells and their activities measured, it was shown that this conserved domain mediated the DNA-binding ability of the receptor, while the C-terminal region was involved in the binding of the appropriate hormone and the N-terminal region was involved in producing maximal induction of transcription of target genes (see Hollenberg *et al.* 1987 for an example of this type of approach).

Sequence analysis of the DNA-binding domain in a variety of receptors showed that it conformed to a consensus sequence of the type Cys–X_2–Cys–X_{13}–Cys–X_2–Cys–$X_{15,17}$–Cys–X_5–Cys–X_9–Cys–X_2–Cys–X_4–Cys. Like the cysteine-histidine finger described in the previous section, the DNA binding of this element is dependent upon the presence of zinc or a related heavy metal such as cadmium. Moreover, this element can be drawn as two conventional zinc fingers, in which four cysteines replace the two cysteine-two histidine structure of the conventional finger in binding zinc and which are separated by a linker region containing the 15–17 variable amino acids (Fig. 7.15). Such a structure is supported by spectrographic analysis of this region of the receptor, which clearly demonstrates the presence of two zinc atoms each co-ordinated by four cysteines in a tetrahedral array (Freedman *et al.* 1988).

Unlike the cysteine–histidine finger, which is present in multiple copies within the proteins which contain it, the cysteine domain of the steroid receptors is present only once in each receptor. This domain is therefore clearly similar to the cysteine–histidine domain in structure and in its co-ordination of zinc, but is distinct in its lack of histidines

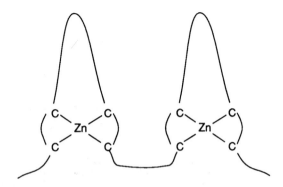

Figure 7.15 Structure of the four-cysteine zinc finger.

Table 7.2

Transcriptional regulatory proteins with multiple cysteine fingers

Finger type	Factor	Species
Cys $_4$ Cys $_5$	Steroid, thyroid receptors	Mammals
Cys $_4$	E1A	Adenovirus
Cys $_6$	GAL4, PPRI, LAC9,	Yeast

and of conserved phenylalanine and leucine residues as well as in its non-repeated structure. The precise evolutionary relationship of these two elements is at present unclear (for further discussion see Frankel & Pabo 1988).

Whatever its precise relationship to the cysteine–histidine finger, it is clear that, like this type of finger, the multi-cysteine domain in the hormone receptors is involved in mediating DNA binding. Similar single domains containing multiple cysteines separated by non-conserved residues have also been identified in other DNA-binding proteins, such as the yeast transcription factors GAL4, PPRI, LAC9, etc., which all contain a cluster of six invariant cysteines, and in the adenovirus transactivator, E1A, which has a cluster of four cysteines within the region that mediates *trans*-activation (Table 7.2; for a review see Evans & Hollenberg 1988).

The existence of a short DNA-binding region in a number of different steroid-receptor proteins which bind distinct but related sequences (Ch. 6, Section 6.2.3) has allowed a dissection of the elements in this structure that are important in sequence-specific DNA binding (reviewed by Berg 1989). Thus, as illustrated in Table 6.4, the sequences that confer responsiveness to glucocorticoid or oestrogen treatment are distinct but related to one another. If the cysteine-rich region of the oestrogen receptor is replaced by that of the glucocorticoid receptor, a chimaeric receptor is obtained which has the DNA-binding specificity of the glucocorticoid receptor but, because all the other regions of the protein are derived from the oestrogen receptor, it continues to bind oestrogen. Hence this hybrid receptor induces the expression of glucocorticoid-responsive genes (which carry its DNA-binding site) in response to treatment with oestrogen (to which it binds) (Green & Chambon 1987; Fig. 7.16). Further so-called finger-swop experiments using smaller parts of this region have shown that this change in specificity can also be achieved by the exchange of the N-terminal, four-cysteine finger, together with the region immediately

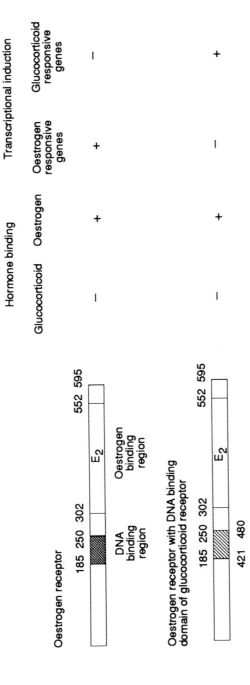

Figure 7.16 Effect of exchanging the DNA-binding domain (shaded) of the oestrogen receptor with that of the glucocorticoid receptor on the binding of hormone and gene induction by the hybrid receptor.

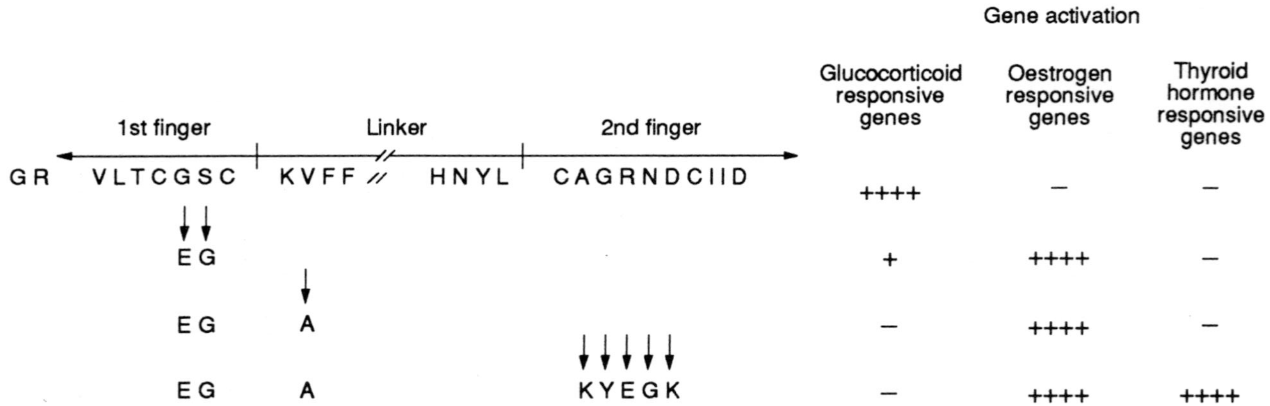

Gene activation

1st finger   Linker   2nd finger

GR VLTCGSC | KVFF // HNYL | CAGRNDCIID

Substitution	Glucocorticoid responsive genes	Oestrogen responsive genes	Thyroid hormone responsive genes
EG	++++	−	−
EG ... A	+	++++	−
EG ... A	−	++++	−
EG ... A ... KYEGK	−	++++	++++

Figure 7.17 Effect of amino-acid substitutions in the zinc finger region of the glucocorticoid receptor on its ability to bind to and activate genes which are normally responsive to different steroid hormones.

following it, which are therefore critical for determining the sequence-specific binding of the DNA.

More recently, these findings have been refined still further by exchanging individual amino acids in this region of the glucocorticoid receptor for their equivalents in the oestrogen receptor. As shown in Fig. 7.17, the alteration of the two amino acids between the third and fourth cysteines of the N-terminal finger to their oestrogen receptor equivalents results in a glucocorticoid receptor which switches on oestrogen-responsive genes. Hence the change of only two critical amino acids within a protein of 777 amino acids can completely change the DNA-binding specificity of the receptor.

The specificity of the hybrid receptor for oestrogen-responsive genes can be further enhanced by changing another amino acid, which is located in the linker region between the two fingers (Fig. 7.17), indicating that this region also plays a role in controlling the specificity of binding to DNA. Interestingly, this region following the finger can form an α-helical structure similar to the recognition helix seen in the helix-turn-helix motif. Thus, the DNA-binding specificity of the steroid receptors appears to involve the co-operation of a zinc finger motif and an adjacent helical motif.

In contrast to the effect of mutations in the first finger and adjacent region, further alteration of five amino acids in the second finger is sufficient to change the binding specificity of the receptor such that it now recognizes the thyroid-hormone-receptor binding sites (Umesono & Evans 1989; Fig. 7.17). Since thyroid-hormone-binding sites do not differ from those of the oestrogen receptor in sequence but only in the spacing between the two halves of the palindromic DNA recognition sequence (Table 6.4), this indicates that the second finger is critical for mediating protein–protein interactions between the two copies of the receptor that bind to the two halves of the palindromic sequence (see Section 6.2.3), and thus for controlling the optimal spacing of these halves for binding of the particular receptor.

Hence by studying the multiple related steroid receptors and their relationship with the related DNA sequences to which they bind, it has been possible to determine the critical role of both the first zinc finger and its adjacent helix in controlling the sequence to which these receptors bind, and of the second zinc finger in determining the spacing of adjacent sequences which is optimal for the binding of each receptor. Once again, the steroid-responsive genes are leading the way in the study of gene regulation.

identical factors, it is possible to envisage the formation of a heterodimer between two different factors, which might have different properties in terms of sequence-specific binding and gene activation compared to homodimers of one or other of the two factors.

An example of this type is seen in the case of the related oncoproteins Fos and Jun. Thus Jun can bind as a homodimer to the AP1 recognition sequence, TGAGTCAG, which mediates transcriptional induction by phorbol esters (Table 6.3). In contrast, Fos cannot bind to DNA alone but can form a heterodimer with the Jun protein. This heterodimer binds to the AP1 recognition site with a thirtyfold greater affinity than the Jun homodimer, and is considerably more effective in enhancing transcription of genes containing the binding site. Both hetero- and homodimer formation and DNA binding are dependent on the leucine zipper motif which is found in both proteins (Kouzarides & Ziff 1988, Turner & Tjian 1989). Hence dimerization by the leucine zipper motif allows two different complexes with different binding affinities and different activity to form on the identical DNA-binding site (Fig. 7.21).

The failure of the Fos protein to form homodimers and its inability to bind to DNA in the absence of Jun has been shown recently to be due to differences in its leucine zipper region from that of Jun. Thus if the leucine zipper region of Fos is replaced by that of Jun the resulting protein can dimerize. This dimerization allows the chimaeric protein to bind to DNA through the basic region of Fos (Neuberg *et al.* 1989),

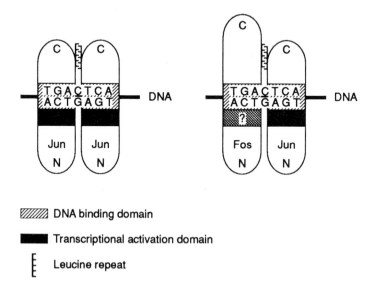

🖎 DNA binding domain

■ Transcriptional activation domain

E Leucine repeat

Figure 7.21 Model for DNA binding by the Jun homodimer and the Fos–Jun heterodimer.

206

which is therefore a fully functional DNA-binding domain.

It is clear, therefore, that the ability of the leucine zipper to facilitate the formation of different dimeric complexes between different transcription factors is likely to play a crucial role in the regulation of gene expression, by producing complexes with different binding affinities and possibly recognizing different sequences from those of the parental homodimers.

7.2.5   Other DNA-binding domains

As more and more genes encoding the transcription factors are cloned and analysed, other DNA-binding domains distinct from those discussed above are being identified. For example, two proteins that bind to the immunoglobulin enhancer have both been shown to contain a DNA-binding domain consisting of two α-helical regions separated by an intervening non-helical loop (Murre *et al.* 1989). This helix-loop-helix motif is distinct from the helix-turn-helix motif discussed above (Section 7.2.2) in having a number of hydrophobic residues located on one side of each helix. Following its identification in these proteins, the helix-loop-helix motif has also been identified in other regulatory proteins, such as those encoded by the *myc* oncogene family and the product of the *Drosophila daughterless* gene. Hence it represents another DNA-binding motif common to a number of regulatory proteins.

Additionally, the DNA-binding regions of transcription factors such as AP2, the serum response factor, and CTF/NF1 have been shown to be distinct from the previously described DNA-binding motifs as well as from each other, and each of these may therefore be the founder member of a new family of DNA-binding elements.

7.3   ACTIVATION OF TRANSCRIPTION

7.3.1   Introduction

Although binding to DNA is a necessary prerequisite for the activation of transcription, it is clearly not in itself sufficient for this to occur. Following binding, the bound transcription factor must somehow increase transcription, either by directly activating the RNA polymerase itself or by facilitating the binding of other transcription factors and the assembly of a stable transcriptional complex. This section will discuss the features of transcription factors that produce this activation and the manner in which they do so (for a review see Ptashne 1988).

7.3.2 Activation domains

IDENTIFICATION OF ACTIVATION DOMAINS

It is clear from the preceding sections of this chapter that transcription factors have a modular structure in which a particular region of the protein mediates DNA binding while another may mediate binding of a co-factor, such as a hormone, and so on. It seems likely, therefore, that a specific region of each individual transcription factor will be involved in its ability to up-regulate transcription following DNA binding.

In the majority of cases, it is clear that such activation regions are distinct from those which produce DNA binding. This domain-type structure is seen clearly in the yeast transcription factor GCN4, which mediates the induction of the genes encoding the enzymes of amino-acid biosynthesis in response to amino-acid starvation. Thus, if a 60 amino-acid region of this protein, containing the DNA-binding region, is introduced into cells, it can bind to the DNA of GCN4-responsive genes but fails to activate transcription (Hope & Stuhl 1986). Hence, although DNA binding is necessary for transcriptional activation to occur, it is not sufficient, and gene activation must be dependent upon a region of the protein that is distinct from that mediating DNA binding.

Unlike the DNA-binding regions, the region of a transcription factor that mediates gene activation cannot, therefore, be identified on the basis of a simple assay of, for example, the ability to bind to DNA or another protein. Rather, a functional assay of gene activation following binding to DNA is required. Activation regions have therefore been identified on the basis of so-called 'domain-swap' experiments, in which the DNA-binding region of one transcription factor is combined with various regions of another factor and the ability to activate transcription of a gene containing the binding site of the first factor is assessed (Fig. 7.22). Following binding of the hybrid factor to the target gene binding site, gene activation will occur only if the hybrid factor also contains an activation domain provided by the second factor, and hence the activation domain can be identified.

Thus, in the case of the yeast transcription factor GCN4 discussed above, if a 60 amino-acid region, outside the DNA-binding domain, is linked to the DNA-binding region of the bacterial regulatory protein, Lex A, the hybrid factor will activate transcription in yeast from a gene containing the binding site for Lex A, whereas neither the Lex A DNA-binding domain nor the GCN4 region will do so alone. Hence this region of GCN4 contains an activation domain, which can increase transcription following DNA binding and is separate from the region of

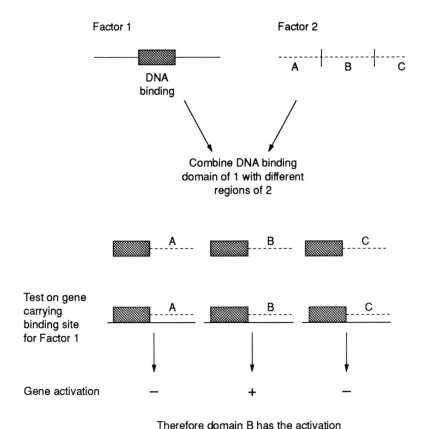

Figure 7.22 Domain swapping experiment, in which the activation domain of factor 2 is mapped by combining different regions of factor 2 with the DNA-binding domain of factor 1 and assaying the hybrid proteins for the ability to activate transcription of a gene containing the DNA-binding site of factor 1.

the protein that normally mediates DNA binding (Hope & Struhl 1986).

Following its initial use in yeast, similar domain-swopping experiments have also been used to identify the activation domains of mammalian transcription factors. In the glucocorticoid receptor, for example, a 200 amino-acid region at the N terminus (amino acids 200–400, see Fig. 7.23), able to produce gene activation, was identified by fusing different regions of the protein to the Lex A DNA-binding domain (Godowski *et al.* 1988). Similarly, using a slightly different approach, Hollenberg & Evans (1988) found that both this region and a 30 amino-acid peptide near the C terminus of the molecule (amino acids 526–556, see Fig. 7.23) produce gene activation independently when fused to the GAL4 DNA-binding domain.

Glucocorticoid receptor

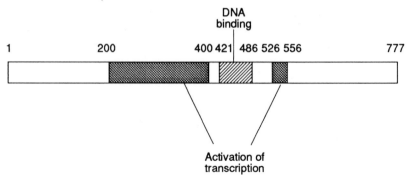

Figure 7.23 Structure of the glucocorticoid receptor, indicating the two regions that can mediate transcriptional activation.

Although other members of the steroid–thyroid hormone receptor family also possess two distinct activation domains, their relative importance in producing gene activation has been shown to vary in different receptors and is also dependent on the target gene that is being activated (reviewed by Green & Chambon 1988).

The success of domain-swop experiments is further proof of the modular nature of transcription factors, allowing the DNA-binding domain of one factor and the activation domain of another to co-operate together to produce gene activation. This is analogous to the exchange of the DNA-binding domain of the glucocorticoid and oestrogen receptors, which allows the creation of a hybrid receptor that binds to oestrogen-responsive genes through the DNA-binding domain but is responsive to the presence of glucocorticoid through the steroid-binding domain of the protein.

An extreme example of this modularity is provided by the herpes simplex virus *trans*-activating protein VP16, which activates transcriptionally the viral immediate-early genes during lytic infection of

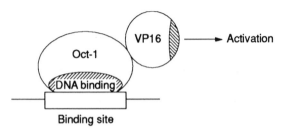

Figure 7.24 Activation of gene transcription by interaction of the cellular factor Oct-1, which contains a DNA-binding domain, and the herpes simplex virus VP16 protein, which contains an activation domain but cannot bind to DNA.

mammalian cells. Thus, although this protein contains a very potent activation region which can strongly induce gene transcription when fused to the DNA-binding domain of GAL4 (Sadowski *et al.* 1988), it contains no DNA binding domain and cannot bind to DNA itself. Rather, following infection, it forms a complex with the cellular octamer-binding protein, Oct-1 (see Section 7.2.2). Oct-1 provides the DNA-binding domain which allows binding to the sequence TAATGARAT (R = purine) in the viral promoters, and activation is achieved by the activation domain of VP16. Hence, in this instance, DNA binding and activation motifs actually reside on separate molecules (Fig. 7.24).

NATURE OF ACTIVATING REGIONS

Acidic domains Although the activating regions identified in various transcription factors do not show strong amino-acid sequence homology to each other, in many cases they have a very high proportion of acidic amino acids, resulting in a strong net negative change. Thus, in the N-terminal activating region of the glucocorticoid receptor, 17 acidic amino acids are contained in an 82 amino-acid region. Similarly, the activating region of the yeast factor GCN4 contains 17 negative charges in an activating region of only 60 amino acids. This has led to the suggestion that activation regions consist of so-called 'acid blobs' or 'negative noodles' which are necessary for the activation of transcription (for a review see Sigler 1988).

In agreement with this idea, mutations in the yeast transcription factor GAL4, which increase the net negative charge on its activating region, result in its increased efficiency, whereas those which decrease the net negative charge have the opposite effect. If recombination is used to create a GAL4 protein carrying several mutations which increase negative charge, the effect on activation is additive, a mutant with four more negative charges than the wild type activating transcription ninefold more efficiently than the wild-type activation region.

Although the acidic nature of the region is clearly important for activation, it is unlikely to be the only feature required. Thus studies of many different activation regions have shown that they can form an α-helical structure which is amphipathic, i.e. all the negatively charged residues are displayed along one surface of the helix, allowing them to contact another protein, while the other surface contains only hydrophobic residues (Fig. 7.25). The importance of such a clustering of the negative charges on one side of the helix was confirmed by synthesizing two 15 amino-acid peptides containing the same amino acids but in a different order, such that only one could form an amphipathic helix with all the negative charges on one side. When

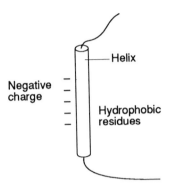

Figure 7.25 Amphipathic helix, in which all the negatively charged amino acids are displayed on one side of the helix.

each of these was linked to the DNA-binding domain of GAL4, only the amphipathic helix-forming peptide could activate transcription, indicating that an activating region requires an amphipathic helix and that a preponderance of randomly arrayed negatively charged amino acids is insufficient (Fig. 7.26, Giniger & Ptashne 1987).

Other activating domains Although acidic domains of the type discussed above have been identified in a wide range of transcriptional activators from yeast to man, it is unlikely that this type of structure is the only one which can mediate transcriptional activation. Indeed, Courey & Tjian (1988) identified two regions of the human Sp1 transcription factor that could mediate activation of transcription, and neither of these was particularly acidic. Instead, each of these two domains was particularly rich in glutamine residues, and the intactness of the glutamine-rich region was essential for transcriptional activation. Similar sequences have also been identified in the homeotic proteins Antennapedia and Cut, in Zeste, another *Drosophila* transcriptional regulator, and in the POU proteins Oct-1 and Oct-2, suggesting that this type of activating region may not be confined to a single protein.

A further type of activation domain has been identified in the transcription factor CTF/NF1 which binds to the CCAAT box present in many eukaryotic promoters (Table 6.2). The activation domain of this protein is not rich in acidic or glutamine residues but instead contains numerous proline residues, forming approximately one-quarter of the amino acids in this region (Mermod *et al.* 1989). Similar proline-rich regions are found in other transcription factors, such as AP2 and Jun, suggesting that, as with glutamine-rich domains, this element is not confined to a single protein.

Hence, as with DNA-binding domains, it is clear that several distinct protein motifs are involved in the activation of transcription.

Units of β
galactosidase

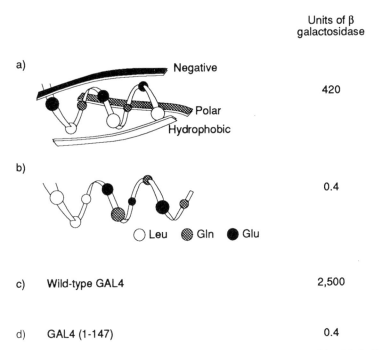

a)   Negative   420

                                      Polar
                                      Hydrophobic

b)   0.4

        ◯ Leu   ◓ Gln   ● Glu

c)   Wild-type GAL4   2,500

d)   GAL4 (1-147)   0.4

Figure 7.26 Effects on gene expression of a sequence forming an amphipathic helix (a) and a sequence of identical amino-acid composition which cannot form an amphipathic helix (b). Gene expression is measured as the amount of β-galactosidase produced by a hybrid β-galactosidase gene containing a binding site for GAL4 in its promoter. Note that the intact GAL4 protein (c) and the truncated GAL4 protein linked to an amphipathic helix (a) can induce gene expression, whereas truncated GAL4 linked to a non-amphipathic helix (b) or alone (d) cannot do so.

7.3.3 How is transcription activated?

The widespread interchangeability of activation domains from yeast, *Drosophila*, and mammalian transcription factors, discussed in the previous section, suggests that one common mechanism may mediate transcriptional activation in a wide range of organisms. This is supported by the observations that mammalian transcription factors, such as the glucocorticoid receptor, can activate a gene carrying their appropriate DNA-binding site when introduced into yeast cells, while the yeast factor GAL4, which normally activates the transcription of yeast genes in the presence of galactose, can activate a gene carrying its binding site in cells of *Drosophila*, tobacco plants, and mammals (reviewed by Guarente 1988, Ptashne 1988). Indeed, yeast and mammalian factors can co-operate together in gene activation, a gene bearing binding sites for GAL4 and the glucocorticoid receptor being synergistically activated by the two factors in mammalian cells.

213

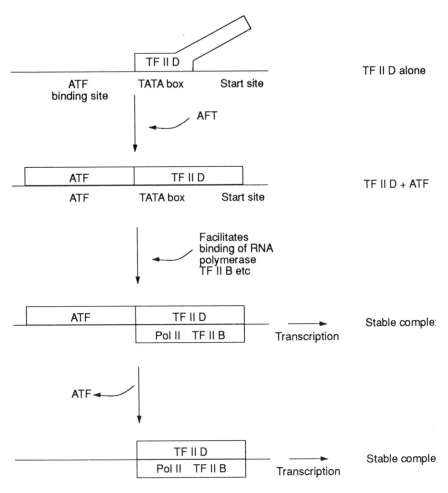

Figure 7.29 Binding of ATF to its binding site alters the binding of TFIID and facilitates the assembly of a stable transcription complex.

to activate transcription and is not simply a consequence of DNA binding.

Similar direct interaction with TFIID has also been observed for the mammalian transcription factor ATF, which mediates the increased transcription of several cellular genes following cyclic AMP treatment, suggesting that this is a general mechanism by which activation domains function (Horikoshi *et al.* 1988b).

The activity of the activation domain in facilitating the binding of TFIID to the TATA box has been shown, in turn, to promote the binding of other components of the transcription complex, such as the RNA polymerase itself and transcription factors TFIIC and TFIIE. Thus, in the ATF case, these factors will complex with the DNA only after ATF and TFIID have both bound to their respective binding sites. The binding of these other factors results in the formation of a stable transcription complex (Fig. 7.29). Once this complex has formed, ATF can be removed without affecting transcription (Hai *et al.* 1988) confirming that its role is in the assembly of the complex rather than in its maintenance.

It is likely, therefore, that many transcription factors involved in tissue-specific gene regulation will act by facilitating the binding of core factors such as RNA polymerase or TFIID, and hence allowing high-level transcription to occur. This role is functionally similar to that of transcription factor TFIIIA in the transcription of the 5S ribosomal RNA by RNA polymerase III, where binding of this factor to the internal control region of the gene facilitated binding of other factors and the assembly of a stable transcriptional complex (Section 6.5.2).

Further studies of the type carried out so far in the ATF system should allow a precise definition of the manner in which activation domains interact with TFIID and the way in which this enhances assembly of the transcriptional complex.

7.4 WHAT ACTIVATES THE ACTIVATORS?

7.4.1 *Introduction*

The crucial role of transcription factors in tissue-specific gene regulation and in the regulation of transcription in response to specific stimuli clearly leads to the question of how such factors are themselves regulated so that they produce gene activation only in a specific tissue or in response to a particular signal.

As discussed in Chapter 6 (Section 6.1.3), the Britten and Davidson model envisaged that the products of regulatory or integrator genes which controlled the activity of other genes would be synthesized in response to a particular signal, or in a particular tissue, and would be absent in other tissues. Hence the activity of the regulated genes would be directly correlated with the presence of the regulatory protein (Fig. 7.30a).

An example of this type is provided by the octamer-binding protein Oct-2 (see Section 7.2.2), which is involved in the stimulation of immunoglobulin gene expression in B-cells. In agreement with this, the Oct-2 protein and the RNA encoding it are present in B-cells but are absent in other cell types, such as HeLa cells, which do not express the immunoglobulin genes. Moreover, artificial expression of the gene encoding Oct-2 in HeLa cells results in the transcription of immuno-

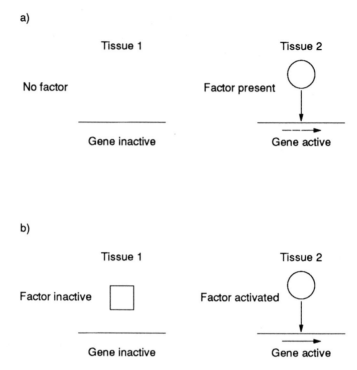

Figure 7.30 Gene activation mediated by the synthesis of a transcription factor only in a specific tissue (a) or its activation in a specific tissue (b).

globulin genes introduced into these cells, confirming that Oct-2 induces transcription of immunoglobulin genes in tissues that contain it (Muller *et al.* 1988).

A contrasting example is provided, however, by another factor involved in immunoglobulin gene transcription, namely NF κB. This factor is detected in a form capable of stimulating immunoglobulin gene transcription only in extracts of mature immunoglobulin-producing B-cells, no activity being detectable in immature B-cells or in non-B-cells, such as HeLa cells. The NF κB protein and its corresponding RNA are detectable in all cell types, however, suggesting that this protein exists in an inactive form in most cell types and is activated in mature B-cells. In agreement with this, active NF κB capable of stimulating immunoglobulin gene transcription can be induced in other cell types, such as T-cells or HeLa cells, by treatment with phorbol esters. This effect occurs even under conditions where new protein synthesis cannot occur, indicating that it takes place via the activation of pre-existing NF κB protein (reviewed by Lenardo & Baltimore 1989).

These two examples illustrate, therefore, how the ability of transcription factors to act only in a particular tissue can be controlled either by tissue-specific synthesis (Fig. 7.30a) or by tissue-specific activation of pre-existing protein (Fig. 7.30b). These two mechanisms will now be considered.

7.4.2 Regulated synthesis of transcription factors

Clearly, all the various levels of gene regulation such as transcription, splicing, translation, etc., which were discussed in Chapter 3 (Section 3.1), could be used to regulate the expression of the genes encoding transcription factors, and there is evidence that several of these are used. These will be discussed in turn with illustrative examples.

REGULATION OF TRANSCRIPTION

Although the low abundance of many transcription factors makes the demonstration of transcriptional control difficult, it has been demon-strated in the case of the C/EBP protein which, as previously described (Section 7.2.4), regulates the transcription of several different liver-specific genes, such as transthyretin and α-1-antitrypsin. This trans-cription factor is made at high level only in the liver and, in agreement with this, significant transcription of its gene is detectable only in this tissue (Xanthopoulos *et al.* 1989). Hence the regulated transcription of the C/EBP gene controls the production of the corresponding protein which, in turn, directly controls the liver-specific transcription of other genes.

REGULATION OF SPLICING

As discussed in Chapter 4 (Section 4.2.3), in mammals alternative splicing is used widely to generate two different forms of a protein, with different functions, from a single gene. A similar theme is seen in the transcription factors, and has been well characterized in the c-*erbA* α gene, which encodes a receptor mediating gene expression in response to thyroid hormone. This protein is therefore a member of the steroid–thyroid hormone receptor gene family discussed earlier (Sec-tion 7.2.3) and, in the presence of thyroid hormone, induces the expression of specific genes. Such induction is dependent upon the presence of a region in the receptor capable of binding thyroid hormone (see below, Section 7.4.3). Interestingly, however, two alternative forms of the protein exist which are encoded by alternative-ly spliced mRNAs. One of these (α-1) contains the hormone-binding domain and can mediate gene activation in response to thyroid hormone, while the other form (α-2) contains another protein sequence instead of part of this domain (Fig. 7.31a). Therefore the α-2 form cannot respond to the hormone, although since it contains the DNA-

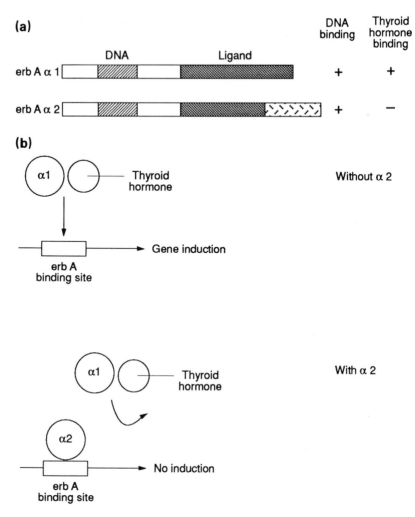

Figure 7.31 a) Relationship of the ErbA α-1 and α-2 proteins. Note that only the α-1 protein has a functional thyroid hormone binding domain. (b) Inhibition of ErbA α-1 binding and of gene activation in the presence of the α-2 protein.

binding domain, it can bind to the binding site for the receptor in hormone-responsive genes. By doing so, it prevents binding of the α-1 form and hence the induction of the gene in response to thyroid hormone (Fig. 7.31b, Koenig *et al*. 1989). Hence these two alternatively spliced forms of the transcription factor, which are made in different amounts in different tissues, mediate opposing effects on thyroid hormone dependent gene expression.

A similar alternative splicing event in the gene encoding a transcription factor has also been observed in the case of the *era-1* gene, which mediates the response of early embryonic cells to treatment with retinoic acid. In this case, however, the alternative splice results in the presence or absence of the region of the gene encoding the homeobox which mediates the DNA-binding activity of the protein (Larosa & Gudas 1988).

REGULATION OF TRANSLATION

As discussed in Chapter 4 (Section 4.5.2), the yeast transcription factor GCN4, which induces the expression of genes involved in amino-acid biosynthesis in response to amino-acid starvation, is itself regulated at the level of translation (reviewed by Fink 1986). Thus, when amino acids are lacking, translation is initiated preferentially at the start point for GCN4 production, whereas when amino acids are abundant translation initiates at short open reading frames upstream of the GCN4 start site and does not reinitiate at the GCN4 site (Fig. 4.19). Hence GCN4 is synthesized in response to amino-acid starvation and activates the genes encoding the enzymes required for the biosynthetic pathways necessary to make good this deficiency.

7.4.3 Regulated activity of transcription factors

Although, as discussed above, many transcription factors are regulated by controlling their synthesis, a number are controlled by regulating the activity of the protein, which is present in many different cell types. Such a system has obvious advantages in that it allows a direct effect of the agent inducing gene expression on the activity of the factor, either by binding to it or by modifying the protein, for example by phosphorylation, and hence results in a rapid response. Examples of these kinds of modifications will now be discussed.

ACTIVATION BY PROTEIN–LIGAND OR PROTEIN–PROTEIN INTERACTION

A simple example of a protein-ligand interaction activating a transcription factor is provided by the ACE1 factor in yeast, which mediates the induction of the metallothionein gene in response to copper. This protein has been shown to undergo a major conformational change upon binding of copper, which allows it to bind to metallothionein gene regulatory sites and induce transcription (Furst *et al.* 1988). Hence the activity of this transcription factor is directly modulated by copper, allowing it to mediate gene activation in response to the metal (Fig. 7.32).

In mammalian cells, the binding of hormone receptors of the steroid–thyroid hormone family to hormone-responsive genes and the induction of their transcription is dependent upon the binding of the

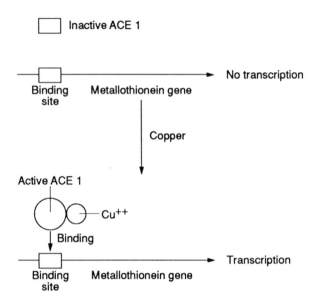

Figure 7.32 Activation of the ACE1 factor in response to copper results in transcription of the metallothionein gene.

appropriate hormone to the receptor, and a hormone-binding region at the carboxyl terminus of these receptors has been identified (Fig. 7.14; reviewed by Beato 1989). Originally it was thought that, as with ACE1, binding of hormone to the receptor activated its ability to bind to DNA and switch on transcription of hormone-responsive genes. However, more recently it has been shown that although the receptor binds to DNA only in the presence of the hormone in the cell, in the test-tube it will bind even when no hormone is present. This has led to the idea that in the cell the receptor is prevented from binding to DNA by its association with another protein, and that the hormone acts to release it from this association and allow it to fulfil its inherent ability to bind to DNA.

In agreement with this idea, the glucocorticoid receptor and other members of the family have been shown to be associated in the cytoplasm with a 90 000 molecular weight heat-inducible protein (hsp90). Upon steroid binding the receptor dissociates from the hsp90 and moves to the nucleus, where it activates gene transcription (Fig. 7.33). Hence the transcription factor is activated by hormone not by a protein–ligand interaction but by disruption of a protein–protein interaction which inhibits the inherent DNA-binding ability of the receptor.

A similar protein–protein interaction also inhibits the activity of the yeast GAL4 transcription factor in the absence of galactose. This

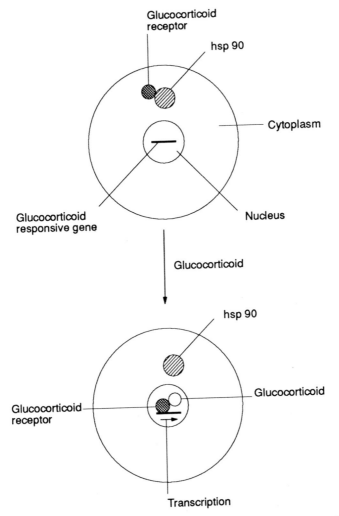

Figure 7.33 Binding of glucocorticoid to the glucocorticoid receptor results in its dissociation from hsp90 and movement of the hormone–receptor complex to the nucleus where it activates the transcription of glucocorticoid-responsive genes.

interaction, however, still allows the factor to bind to DNA but its acidic activating region is masked by another protein, GAL80. In the presence of galactose, GAL80 dissociates and activation can occur (Fig. 7.34). Interestingly, if an acidic region is linked artifically to GAL80, it will activate transcription through the DNA-bound GAL4 protein in the same way as the herpes simplex virion protein VP16 acts through bound octamer binding protein (see Section 7.3.2).

223

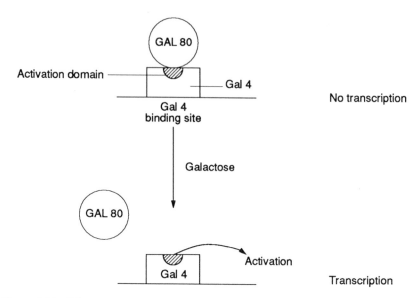

Figure 7.34. Galactose activates gene expression by removing the GAL80 protein from the DNA-bound GAL4 protein, unmasking the activation domain on GAL4.

ACTIVATION BY PROTEIN MODIFICATION

Many transcription factors are modified extensively by the addition, for example, of 0-linked monosaccharide residues or by phosphorylation. Such modifications represent obvious targets for agents that induce gene activation. Thus, such agents could act by altering the activity of a modifying enzyme, such as a kinase, which in turn would act on the transcription factor, resulting in its activation and the switching on of gene expression.

The best-characterized example of this type involves the mammalian CREB factor, which mediates the induction of several genes in response to treatment with cyclic AMP and binds to the cyclic AMP response element in these genes (Table 6.3). The dimerization of CREB and its ability to induce transcription is enhanced greatly by its phosphorylation at several sites in the protein (Yamamoto *et al.* 1988). Hence the activation of gene expression by cyclic AMP is likely to be mediated by its activation of a protein kinase which, in turn, phosphorylates and activates CREB (Fig. 7.35).

The regulation of phosphorylation is also likely to be involved in the activation of NF κB in response to phorbol ester treatment, as described above (Section 7.4.1). In this case, however, the target for phosphorylation is not the transcription factor itself but rather an associated protein, I κB, which inhibits NF κB activity. Prior to phorbol ester treatment, NF κB is bound to I κB as an inactive complex which is

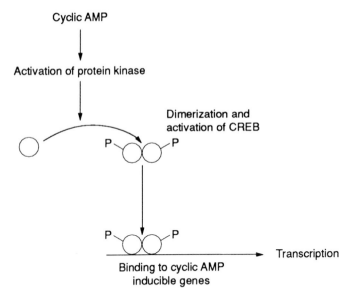

Figure 7.35 Phosphorylation of the CREB transcription factor in response to cyclic AMP leads to dimerization, which allows DNA binding and the activation of cyclic AMP inducible genes.

located in the cytoplasm. Following treatment, however, the complex dissociates and NF κB moves to the nucleus where it activates gene expression (reviewed by Lenardo & Baltimore 1989). Such activation is clearly similar to the dissociation of the steroid receptor from hsp90 following steroid treatment, although in this case phosphorylation rather than steroid binding dissociates the protein–protein complex. This example therefore links the two types of activation processes we have discussed, involving both protein–protein interaction and reversible protein modification.

7.5 CONCLUSIONS

Although the existence of transcription factors had been known or inferred for some time, it is clear that the cloning of the genes encoding these factors has increased our knowledge of their functional properties greatly. A number of different motifs which mediate DNA binding, protein dimerization, and transcriptional activation have been identified and these are listed in Table 7.3. The list is likely to grow, however, as more factors are cloned and characterized.

Further information is also likely to become available about the manner in which transcription factors become activated in particular

Gehring, W. J. 1987. Homeoboxes in the study of development. *Science* **236**, 1245–52.

Giniger, E. & M. Ptashne 1987. Transcription in yeast activated by a putative amphipathic alpha helix linked to a DNA binding unit. *Nature* **330**, 670–2.

Godowski, P. J., D. Picard & K. Yamamoto 1988. Signal transduction and transcriptional regulation by glucocorticoid receptor-Lex A fusion proteins. *Science* **241**, 812–16.

Green, S. & P. Chambon 1987. Oestradiol induction of a glucocorticoid-response gene by a chimaeric receptor. *Nature* **325**, 75–8.

Green, S. & P. Chambon 1988. Nuclear receptors enhance our understanding of transcription regulation. *Trends in Genetics* **4**, 309–14.

Guarente, L. 1988. UASs and enhancers: common mechanism of transcriptional activation in yeast and mammals. *Cell* **52**, 303–5.

Hai, T., M. Horikoshi, R. G. Roeder & M. R. Green 1988. Analysis of the role of the transcription factor ATF in the assembly of a functional preinitiation complex. *Cell* **54**, 1043–51.

Hanes, S. D. & R. Brent 1989. DNA specificity of the bicoid activator protein is determined by homeodomain recognition helix 9. *Cell* **57**, 1275–83.

He, X., M. N. Treacy, D. M. Simmons, H. A. Ingraham, L. S. Swanson & M. G. Rosenfeld 1989. Expression of a large family of POU-domain regulatory genes in mammalian brain development. *Nature* **340**, 35–42.

Herr, W., R. A. Sturm, R. G. Clerc, L. M. Corcoran, D. Baltimore, P. A. Sharp, H. A. Ingraham, M. G. Rosenfeld, M. Finney, G. Ruvkun & H. R. Horvitz 1988. The POU domain: a large conserved region in the mammalian pit-1, oct-1, oct-2 and *Caenorhabditis elegans* unc-86 gene products. *Genes and Development* **2**, 1513–16.

Hollenberg, S. M. & R. M. Evans 1988. Multiple and cooperative trans-activation domains of the human glucocorticoid receptor. *Cell* **55**, 899–906.

Hollenberg, S. M., V. Giguere, P. Segui & R. M. Evans 1987. Colocalization of DNA-binding and transcriptional activation functions in the human glucocorticoid receptor. *Cell* **49**, 39–46.

Hope, I. A. & K. Struhl 1986. Functional dissection of a eukaryotic transcriptional activator GCN4 of yeast. *Cell* **46**, 885–94.

Horikoshi, M., M. F. Carey, H. Kakidani & R. G. Roeder 1988a. Mechanism of action of a yeast activator: direct effect of GAL4 derivatives on mammalian TFIID-promoter interactions. *Cell* **54**, 665–9.

Horikoshi, M., T. Hai, Y.-S. Lin, M. R. Green & R. G. Roeder 1988b. Transcription factor ATF interacts with the TATA box factor to facilitate establishment of a preinitiation complex. *Cell* **54**, 1033–42.

Ingham, P. W. 1988. The molecular genetics of embryonic pattern formation in *Drosophila*, *Nature* **335**, 25–34.

Ingraham, H. A., R. Chen, H. J. Mangalam, H. P. Elsholtz, S. C. Flynn, C. R. Linn, D. M. Simmons, L. Swanson & M. G. Rosenfeld 1988. A tissue specific factor containing a homeo domain specifies a pituitary phenotype. *Cell* **55**, 519–29.

Jaynes, J. B. & P. H. O'Farrell 1988. Activation and repression of transcription by homeodomain-containing proteins that bind a common site. *Nature* **336**, 744–9.

Johnson, P. F. & S. L. McKnight 1989. Eukaryotic transcriptional regulatory proteins. *Annual Review of Biochemistry* **58**, 799–839.

Kadonga, J. T. & R. Tjian 1986. Affinity purification of sequence-specific DNA binding proteins. *Proceedings of the National Academy of Sciences of the USA* **83**, 5889–93.

Kadonga, J. T., K. R. Carner, F. R. Masiarz & R. Tjian 1987. Isolation of cDNA encoding the transcription factor Sp1 and functional analysis of the DNA binding domain. *Cell* **51**, 1079–90.

Klug, A. & D. Rhodes 1987. Zinc fingers: a novel protein motif for nucleic acid recognition. *Trends in Biochemical Sciences* **12**, 464–9.

Koenig, R. G., M. A. Lazar, R. A. Hoden, G. A. Brent, P. R. Larsen, W. W. Chin & D. D. Moore 1989. Inhibition of thyroid hormone action by a non-hormone binding c-erbA protein generated by alternative RNA splicing. *Nature* **337**, 659–61.

Kouzarides, T. & E. Ziff 1988. The role of the leucine-zipper in the fos-jun interaction. *Nature* **336**, 646–51.

Krasnow, M. A., E. C. Saffman, K. Kornfeld & D. S. Hogness 1989. Transcriptional activation and repression by ultrabithorax proteins in cultured *Drosophila* cells. *Cell* **57**, 1031–43.

Landschultz, W. H., P. F. Johnson & S. L. McKnight 1988. The leucine zipper: a hypothetical structure common to a new class of DNA binding proteins. *Science* **240**, 1759–64.

Landschultz, W. H., P. F. Johnson & S. L. McKnight 1989. The DNA binding domain of the rat liver nuclear protein C/EBP is bipartite. *Science* **243**, 1681–8.

Larosa, G. J. & L. J. Gudas 1988. Early retinoic acid-induced F9 terato-carcinoma stem cell gene ERA-1: alternative splicing creates transcripts for a homeobox-containing protein and one lacking the homeobox. *Molecular and Cellular Biology* **8**, 3906–17.

Lenardo, M. J. & D. Baltimore 1989. NF-kappa B: a pleotropic mediator of inducible and tissue-specific gene control. *Cell* **58**, 227–9.

Levine, M. & T. Hoey 1988. Homeobox proteins as sequence-specific transcription factors. *Cell* **55**, 537–40.

Mermod, N., E. A. O'Neill, T. J. Kelley & R. Tjian 1989. The proline-rich transcriptional activator of CTF/NF-1 is distinct from the replication and DNA binding domain. *Cell* **58**, 741–53.

Mihara, H. & E. T Kaiser 1988. A chemically synthesized Antennapedia homeo domain binds to a specific DNA sequence. *Science* **242**, 925–7.

Miller, J., A. D. McLachlan & A. Klug 1985. Repetitive zinc-binding domains in the protein transcription factor III A from *Xenopus* oocytes. *EMBO Journal* **4**, 1609–14.

Mitchell, P. J. & R. Tjian 1989. Transcriptional regulation in mammalian cells by sequence specific DNA binding proteins. *Science* **245**, 371–8.

Muller, M., M. Affolter, W. Leupin, G. Otting, K. Wuthrich & U. J. Gehring 1988. Isolation and sequence specific DNA binding of the Antennapedia homeodomain. *EMBO Journal* **7**, 4299–304.

Muller, M. M., S. Ruppert, W. Schaffner & P. Matthias 1988. A cloned octamer transcription factor stimulates transcription from lymphoid-specific pro-moters in non-B cells. *Nature* **336**, 544–51.

Murre, C., P. S. McCaw & D. Baltimore 1989. A new DNA binding and dimerization motif in immunoglobulin enhancer binding, daughterless, MyoD and myc proteins. *Cell* **56**, 777–83.

Neuberg, M., J. Adamkiewicz, J. B. Hunter & R. Muller 1989. A Fos protein containing the Jun leucine zipper forms a homeodimer which binds to the AP1 binding site. *Nature* **341**, 243–5.

Page, D. C., R. Masher, E. M. Simpson, E. M. C. Fisher, G. Mardon, J. Pollack, B. McGillivray, A. deLa Chapelle & L. G. Bram 1987. The sex determining region of the human Y chromosome encodes a finger protein. *Cell* **51**, 1091–104.

causing oncogenes and the processes that led to their discovery (reviewed by Weinberg 1985, Bishop 1987).

As long ago as 1911, Peyton Rous showed that a connective tissue cancer in the chicken was caused by an infectious agent. This agent was subsequently shown to be a virus and was named Rous Sarcoma virus (RSV) after its discoverer. This cancer-causing, tumorigenic virus is a member of a class of viruses called retroviruses, whose genome consists of RNA rather than DNA as in most other organisms.

The majority of viruses of this type do not cause cancers but simply infect a cell and produce a persistent infection with continual production of virus by the infected cell. In the case of RSV, however, such infection also results in the conversion of the cell into a cancer cell, capable of indefinite growth and eventually killing the organism containing it.

In the case of non-tumorigenic retroviruses, the genome contains only three genes, which are known as *gag*, *pol*, and *env* and which function in the normal life cycle of the virus (Fig. 8.1). Following

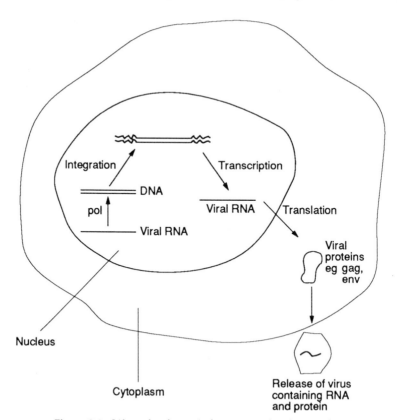

Figure 8.1 Life cycle of a typical non-tumorigenic retrovirus.

232

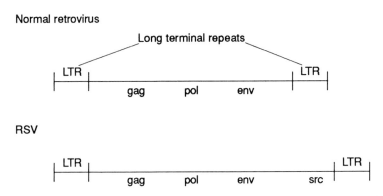

Figure 8.2 Comparison of the genome of a non-tumorigenic retrovirus with that of the tumorigenic retrovirus RSV.

cellular entry, the viral RNA is converted into DNA by the action of the Pol protein, and this DNA molecule then integrates into the host chromosome. Subsequent transcription and translation of this DNA produces the viral structural proteins Gag and Env which coat the viral RNA genome, yielding viral particles that leave the cell to infect other susceptible cells.

Inspection of the RSV genome (Fig. 8.2) reveals an additional gene, known as the *src* gene, which is absent in the other viruses, suggesting that this gene is responsible for the ability of the virus to cause cancer. This idea was confirmed subsequently by showing that if the *src* gene alone was introduced into normal cells it was able to transform them to a cancerous phenotype. This gene was therefore called an oncogene (from the Greek *onkas* for mass or tumour) or cancer-causing gene.

Following the identification of the *src* oncogene in RSV, a number of other oncogenes were identified in other oncogenic retroviruses infecting both chickens and mammals, such as the mouse or rat, and over 20 genes of this type are now known (Table 8.1).

The identification of individual genes that are able to cause cancer obviously opened up many avenues of investigation for the study of this disease. From the point of view of gene regulation, however, the most exciting aspect of oncogenes was provided by the discovery that these cancer-causing genes are derived from genes present in normal cellular DNA. Thus, using Southern blotting techniques (Section 2.2.4) Takeya & Hanafusa (1985) detected a cellular equivalent of the viral *src* gene in the DNA of both normal and cancer cells (Fig. 8.3), and also showed that an mRNA capable of encoding the Src protein was produced in normal cells. Subsequent studies identified cellular equivalents of all the retroviral oncogenes. Many of these cellular equivalents of the viral oncogenes have now been cloned and shown to

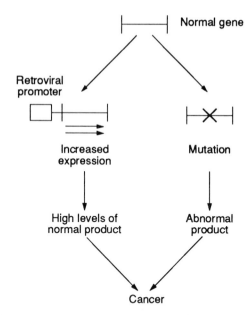

Figure 8.5 A cellular proto-oncogene can be converted into a cancer-causing oncogene by increased expression or by mutation.

alteration by mutation within the cellular genome. Hence these genes play an important role in the generation of human cancer.

The potential risk of such proto-oncogenes causing cancer raises the question of why these genes have not been deleted during evolution. In fact, proto-oncogenes have been highly conserved in evolution, equivalents to mammalian and chicken oncogenes having been found not only in other vertebrates but also in invertebrates, such as *Drosophila*, and even in single-celled organisms, such as yeast (see, for example, De Feo-Jones *et al.* 1983).

The extraordinary evolutionary conservation of many of these genes, despite their potential danger, led to the suggestion that their products were essential for the processes regulating the growth of normal cells, and that their malregulation or mutation therefore results in abnormal growth and cancer. This idea has been confirmed abundantly as more and more proto-oncogenes have been characterized and shown to encode growth factors that stimulate the growth of normal cells (Waterfield *et al.* 1983), cellular receptors for growth factors (Downward *et al.* 1984), and other cellular proteins involved in transmitting the growth signal within the cell, either by acting as a protein kinase enzyme or by binding GTP (McGrath *et al.* 1984).

In the modulation of cellular growth, such proto-oncogene products may interact with the products of other cellular regulatory loci, known as anti-oncogenes. These genes, identified originally on the basis that their mutation or deletion led to the production of a tumour, are believed to have a suppressive effect on cellular growth. Deletion of the best-characterized gene of this type, *Rb-1*, is responsible for the human eye tumour retinoblastoma. Its product has been shown to interact with oncogene proteins and is believed to play a critical role in the regulation of the cell cycle (reviewed by Cooper & Whyte 1989).

Ultimately the growth regulatory pathways controlled by oncogene and anti-oncogene products end in the nucleus, with the activation of genes whose corresponding proteins are required by the growing cell. It is not surprising, therefore, that several proto-oncogenes have been shown to encode transcription factors that regulate the expression of genes activated in growing cells (Table 8.1).

Hence oncogenes present two aspects of importance from the point of view of the regulation of gene expression. First, since cancer is often caused by elevated expression of cellular oncogenes, the processes whereby this occurs are of interest, both from the point of view of the aetiology of cancer and for the light they throw on the mechanisms which regulate gene expression. This topic is therefore discussed in Section 8.3. Secondly, the study of the transcription factors encoded by a few proto-oncogenes has led to a better understanding of the processes regulating gene expression in both cells growing normally and in cancer cells, and this is discussed in Section 8.4.

8.3   ELEVATED EXPRESSION OF ONCOGENES

The products of cellular proto-oncogenes play a critical role in cellular growth control and, in many cases, are synthesized only at specific times and in small amounts. It is not surprising, therefore, that transformation into a cancer cell can result when these genes are expressed at high levels in particular situations.

The simplest example of such overexpression occurs in the case of retroviruses where, as we have already discussed, the oncogene comes under the influence of the strong promoter contained in the retroviral long terminal repeat (LTR) region and is hence expressed at a high level (Fig. 8.5).

A similar up-regulation due to the activity of a retroviral promoter is also seen in the case of avian leukosis virus (ALV) of chickens. Unlike the retroviruses described so far, however, this virus does not carry its own oncogene. Rather it transforms by integrating into cellular DNA next to a cellular oncogene, the c-*myc* gene (Fig. 8.6a). The expression of the c-*myc* gene is brought under the control of the strong promoter in the retroviral LTR, and is hence expressed at levels 20–50 times

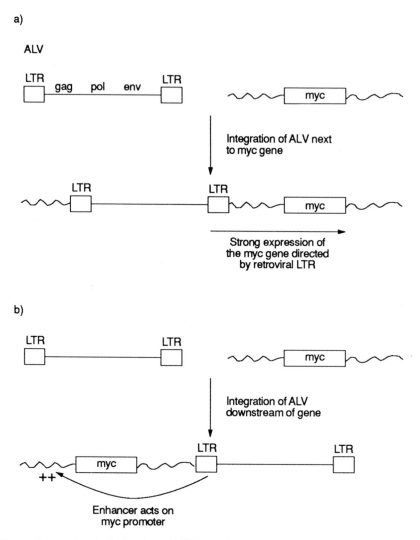

Figure 8.6 Avian leukosis virus (ALV) can increase expression of the *myc* proto-oncogene either via promoter insertion (a) or by the action of its enhancer (b).

higher than normal, producing transformation (Hayward *et al.*, 1981). This process is known as promoter insertion.

In other cases of this type, the ALV virus has been shown to have integrated downstream rather than upstream of the c-*myc* gene. Hence the elevated expression of the c-*myc* gene in these cases cannot be due to promoter insertion. Rather, it involves the action of an enhancer element in the viral LTR which activates the c-*myc* gene's own promoter. As discussed in Chapter 6 (Section 6.3.1), enhancers, unlike

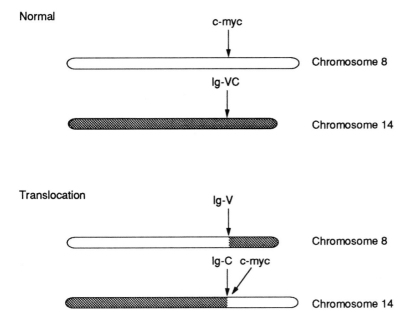

Figure 8.7 Translocation of the c-*myc* gene from chromosome 8 to the immunoglobulin heavy-chain gene locus on chromosome 14 that occurs in cases of Burkitt's lymphoma.

promoters, can act in either orientation and at a distance, and hence the ALV enhancer can activate the *myc* promoter readily from a position downstream of the gene (Fig. 8.6b).

These cases are of interest from the point of view of gene regulation, as indicating how viral regulatory systems can subvert cellular control processes. Of potentially greater interest, however, are the cases where up-regulation of a cellular oncogene occurs through the alteration of internal cellular regulatory processes, rather than through viral intervention. The best-studied example of this type concerns the increased expression of the c-*myc* oncogene which occurs in the transformation of B-cells in cases of Burkitt's lymphoma in humans or in the similar plasmacytomas that occur in mice.

When these tumours were studied, it was noted that they commonly contained very specific chromosomal translocations which involved the exchange of genetic material between chromosome 8 and chromosome 14 (Fig. 8.7; for reviews see Leder *et al.* 1983, Rabbits 1985). Most interestingly, the region of chromosome 8 involved includes the c-*myc* gene, and the translocation results in the *myc* gene being moved to chromosome 14, where it becomes located adjacent to the gene encoding the immunoglobulin heavy chain. This translocation results in the increased expression of the *myc* gene which is observed in the tumour cells.

Normal

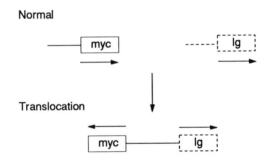

Figure 8.8 Head-to-head orientation of the translocated *myc* gene and the immunoglobulin gene.

Detailed study of the processes mediating such increased expression has indicated that it is produced by different mechanisms in different lymphomas, depending on the precise break points of the translocation within the c-*myc* and immunoglobulin genes (reviewed by Perry 1983, Cory 1986). In all cases studied, however, the break point of the translocation occurs within the immunoglobulin gene, resulting in a truncated gene lacking its promoter being linked to the c-*myc* gene. This fact, together with the fact that the genes are always linked in a head-to-head orientation (Fig. 8.8), indicates that the up-regulation of the c-*myc* gene does not occur via a simple promoter insertion mechanism (see Fig. 8.6a) in which it comes under the control of the immunoglobulin promoter. In some cases, however, the B-cell-specific enhancer element, which is located between the joining and constant regions of the immunoglobulin genes (see Ch. 2, Section 2.4 and Ch. 6, Section 6.3.2), is brought close to the *myc* promoter by the translocation (Fig. 8.9). This enhancer element is highly active in B-cells and can activate the *myc* promoter in a manner analogous to the enhancer of ALV (see Fig. 8.6b).

Hence in this case the c-*myc* gene is up-regulated by the action of the B-cell-specific regulatory mechanisms of the immunoglobulin gene, the immunoglobulin enhancer activating the c-*myc* promoter rather than its own promoter, which has been removed by the translocation. In other cases, however, this does not appear to be the case, the break point of the translocation having removed both the immunoglobulin promoter and enhancer. This leaves the c-*myc* gene adjacent to the constant region of the immunoglobulin gene without any obvious B-cell-specific regulatory elements. In these cases it is likely that the up-regulation of the c-*myc* gene arises from its own truncation in the translocation. This results in the removal of negative regulatory elements that normally repress its expression, such as the upstream silencer element which

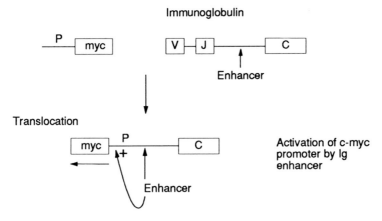

Figure 8.9 In some cases the enhancer of the immunoglobulin heavy-chain gene activates the *myc* gene promoter.

normally represses the c-*myc* promoter (see Ch. 6, Section 6.3.4 and Fig. 8.10).

More extensive truncation of the c-*myc* gene, involving the removal of transcribed sequences rather than upstream elements, has also been observed in some tumours. Frequently, this involves the removal of the first exon of the c-*myc* gene which does not contain any protein-coding information. This exon may thus fulfil a regulatory role by modulating the stability of the c-*myc* RNA or by affecting its translatability. Hence its removal could enhance the level of c-*myc* protein by increasing the stability or the efficiency of translation of the

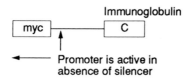

Figure 8.10 Activation of *myc* gene expression can be achieved by removal of the upstream silencer element.

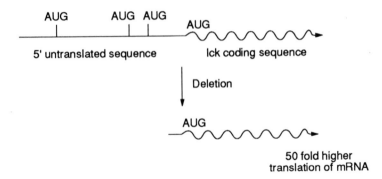

Figure 8.11 Increased translation of the *lck* proto-oncogene mRNA can be achieved by deletion of AUG translational start sites upstream of the AUG that initiates the coding sequence of the Lck protein.

c-*myc* RNA produced at a constant level of transcription. A similar increase in gene expression could be achieved by the removal of sequences within the first intervening sequence which inhibit the transcriptional elongation of the nascent c-*myc* transcript (see Ch. 3, Section 3.3).

The increased expression of an oncogene produced by the removal of sequences that negatively regulate it is also seen in the case of the *lck* proto-oncogene, which encodes a tyrosine kinase related to the c-*src* gene product. In this case, activation of the oncogene in tumours is accompanied by the removal of sequences within its 5′ untranslated region, upstream of the start site of translation (Marth *et al.* 1988). The removal of these sequences results in a fiftyfold increase in the initiation of translation of the *lck* mRNA into protein. Most interestingly, the region removed contains three AUG translation initiation codons which are located upstream of the correct initiation codon for production of the Lck protein (Fig. 8.11). The elimination of these codons results in increased translation initiation from the correct AUG, suggesting that initiation at the upstream codons inhibits correct initiation. This is exactly analogous to the regulation of translation of the GCN4 protein, which was discussed in Chapter 4 (Section 4.5.2).

It is likely that the processes regulating the expression of oncogenes such as c-*myc* and *lck*, which are revealed by studying their overexpression in tumours, also play a role in their normal pattern of regulation during cellular growth. Hence their further study will throw light not only on the mechanisms of tumourgenesis but also on the processes regulating gene expression in normal cells.

Other examples of up-regulation of oncogene expression in tumourgenesis may occur, however, by mechanisms that are unique to the transformed cell. Thus, as discussed in Chapter 2 (Section 2.3) DNA amplification is relatively rare in normal cells. In turmours, however, it is observed frequently for specific oncogenes (Collins & Groudine 1982), and results in the presence of regions of amplified DNA which are visible in the microscope as homogeneously staining regions or as double minute chromosomes. Such amplification is especially common in human lung tumours and brain tumours, and frequently involves the c-*myc* related genes, N-*myc* and L-*myc*. The expression of these genes in the tumour is increased dramatically due to the presence of up to 1000 copies of the gene in the tumour cell.

A variety of mechanisms, involving both the subversion of normal control processes or abnormal events occurring in the tumour cell, therefore result in the observed overexpression of oncogenes in tumour cells. When taken together with the production of abnormal oncogene products due to mutations, it is likely that these processes are involved in the majority of human cancers.

8.4   TRANSCRIPTION FACTORS AS ONCOGENES

As described in Section 8.2, the isolation of the cellular genes encoding particular oncogene products led to the realization that they were involved in many of the processes regulating cellular growth. Ultimately the onset and continuation of cellular growth is likely to involve the activation of cellular genes that are not expressed in quiescent cells. It is not surprising, therefore, that several proto-oncogenes have been shown to encode transcription factors which regulate the transcription of genes activated in growing cells, and several of these cases will be discussed.

8.4.1   Fos, Jun, and AP1 Reviewed by Curran & Franza (1988)

The chicken retrovirus, avian sarcoma virus ASV17, contains an oncogene, v-*jun*, whose equivalent cellular proto-oncogene encodes a nuclearly located DNA-binding protein. Sequence analysis of this protein revealed that it showed significant homology to the DNA-binding domain of the yeast transcription factor GCN4, suggesting that it might bind to similar DNA sequences (Fig. 8.12; Vogt et al. 1987). Interestingly, GCN4 itself had been shown previously to bind to similar sequences to those bound by a factor, AP1 (activator protein 1), which had been detected in mammalian cell extracts by its DNA-binding activity (Fig. 8.13).

This relationship of Jun and AP1 to the sequence and binding activity, respectively, of GCN4 led to the suggestion that Jun might be related to AP1. This was confirmed by the findings that antibody to Jun

Figure 8.12 Comparison of the carboxyl-terminal amino-acid sequences of the chicken Jun protein and the yeast transcription factor GCN4. Boxes indicate identical residues.

reacted with purified AP1 preparations, and that Jun expressed in bacteria was capable of binding to AP1 binding sites in DNA (Bos *et al.* 1988). Moreover, Jun was capable of stimulating transcription from promoters containing AP1 binding sites but not from those which lacked these sites (Angel *et al.* 1988). Hence the *jun* oncogene encodes a sequence-specific DNA-binding protein capable of stimulating transcription of genes containing its binding site, which is identical to the AP1 binding site.

Although Jun undoubtedly binds to AP1-binding sites, preparations of AP1 purified on the basis of this ability contain several other proteins in addition to c-Jun. Several of these are encoded by genes related to *jun*, but another is the product of a different proto-oncogene, namely c-*fos*. Interestingly, however, although Fos is present in AP1 preparations, it does not bind to DNA when present alone but requires the presence of another protein, known as p39, for DNA binding. This p39 protein is the product of the c-*jun* gene (Rauscher *et al.* 1988). Hence, in addition to its ability to bind to AP1 sites alone, Jun can also form a complex with Fos that binds to this site. As discussed in Chapter 7 (Section 7.2.4), this association takes place through the leucine zipper domains of the two proteins and results in a heterodimer complex which binds to the AP1 binding site with much greater affinity than the Jun homodimer (Halazonetis *et al.* 1988).

Hence both Fos and Jun which were identified originally through their association with oncogenic retroviruses, are also cellular transcription factors. Such a finding raises the question of the normal role of these factors and how their incorporation into a retrovirus leads to

DNA binding site

GCN 4	5'	T	G	A	C/G	T	C	A	T	3'
AP-1	5'	T	G	A	G	T	C	A	G	3'

Figure 8.13 Relationship of the DNA-binding sites for the yeast transcription factor GCN4 and the mammalian transcription factor AP1.

cancer. In this regard it is of obvious interest that the AP1-binding site is involved in mediating the induction of genes that contain it in response to treatment with phorbol esters, which are also capable of promoting cancer. Thus not only do many phorbol ester-inducible genes contain AP1 binding sites (see Table 6.3) but, in addition, transfer of AP1 binding sites to a normally non-inducible gene renders that gene inducible by phorbol esters (Angel *et al.* 1987, Lee *et al.* 1987). Increased levels of Jun and Fos are also observed in cells after treatment with phorbol esters (Lamph *et al.* 1988), indicating that these substances act by increasing the levels of Fos and Jun, which in turn cause increased transcription of other genes containing AP1-binding sites which mediate induction by the Fos–Jun complex.

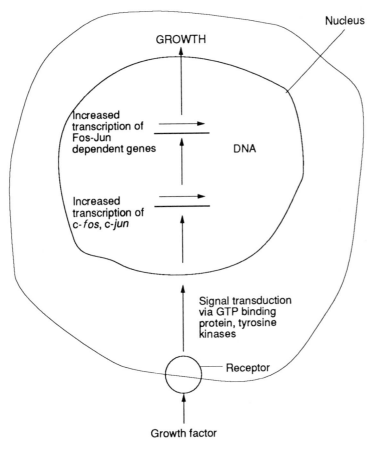

Figure 8.14 Growth factor stimulation of cells results in increased transcription of the c-*fos* and c-*jun* genes, which in turn stimulates transcription of genes which are activated by the Fos–Jun complex.

Most interestingly, increased levels of Jun and Fos are also produced by treatment with serum or growth factors which stimulates the growth of quiescent cells. Hence the transduction of the signal to grow, which begins with the growth factors and their cellular receptors and continues with intra-cytoplasmic signal transducers such as protein kinases and GTP-binding proteins, ends in the nucleus with the increased level of the transcription factors Jun and Fos (Fig. 8.14). These proteins will then activate the genes whose products are necessary for the process of growth itself.

Clearly, it is relatively easy to fit the oncogenic properties of Jun and Fos into this framework. Thus, if these proteins are normally produced in response to growth-inducing signals and activate growth, their continual synthesis following retroviral infection or otherwise will result in a cell which will be stimulated to grow continually and will not respond to growth-regulating signals. Such continuous uncontrolled growth is characteristic of the cancer cell.

In agreement with this idea, mutations in the leucine zipper region of Fos, which abolish its ability to dimerize with Jun and induce genes containing AP1 sites, also abolish its ability to transform cells to a cancerous phenotype (Schuermann et al. 1989). Hence, the ability of Fos to cause cancer is directly linked to its ability to act as a transcription factor for genes containing the appropriate binding site.

8.4.2 v-erbA and the thyroid hormone receptor

Unlike most other retroviruses, avian erythroblastosis virus (AEV) carries two cellular oncogenes, v-erbA and v-erbB (reviewed by Beug et al. 1985). When c-erbA, the cellular equivalent of v-erbA, was cloned, it was shown to encode the cellular receptor that mediates the response to thyroid hormone (Sap et al. 1986, Weinberger et al. 1986).

As discussed in Chapter 7 (Section 7.4.2), this receptor is a member of the super-family of steroid–thyroid hormone receptors which, following binding of a particular hormone, induce the transcription of genes containing a binding site for the hormone–receptor complex. In the case of ErbA, the protein contains a region that can bind thyroid hormone. Following such binding, the hormone–receptor complex induces transcription of thyroid-hormone-responsive genes, such as those encoding growth hormone or the heavy chain of the myosin molecule, by binding to its appropriate DNA recognition site within their promoters (Fig. 8.15).

The finding that the cellular homologue of the v-erbA oncogene is a hormone-responsive transcription factor therefore provides a further connection between oncogenes and cellular transcription factors. It raises the question, however, of the manner in which the transfer of

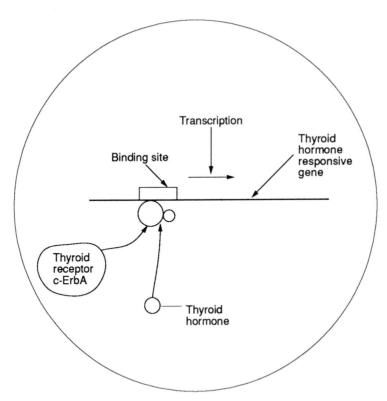

Figure 8.15 The c-*erbA* gene encodes the thyroid hormone receptor and activates transcription in response to thyroid hormone.

the thyroid hormone receptor to a virus can result in transformation. To answer this question it is necessary to compare the protein encoded by the virus with its cellular counterpart. As shown in Figure 8.16 the ErbA protein has the typical structure of a member of the steroid–thyroid hormone family (see also Fig. 7.14), containing both DNA-binding and hormone-binding regions. The viral ErbA protein is generally similar except that it is fused to a portion of the retroviral gag protein at its N terminus. It also contains a number of mutations in both the DNA-binding and hormone-binding regions, as well as a small deletion in the hormone-binding domain.

Of these changes, it is the alterations in the hormone-binding domain which have the most significant effects on the function of the protein and which are thought to be critical for transformation. Thus these changes abolish the ability of the protein to bind thyroid hormone (Sap *et al.* 1986). As such they render this protein functionally analogous to the alternatively spliced form of the c-*erbA* gene product discussed in Chapter 7 (Section 7.4.2), which lacks the hormone-

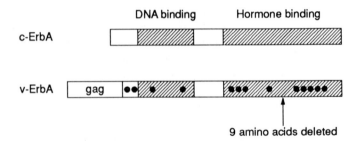

Figure 8.16 Relationship of the cellular ErbA protein and the viral protein. The black dots indicate single amino acid differences between the two proteins while the arrow indicates the region where nine amino acids are deleted in the viral protein.

binding domain and dominantly represses the ability of the hormone-binding receptor form to activate thyroid-hormone-responsive genes.

The idea that the non-hormone-binding viral ErbA protein might also be able to do this has been confirmed recently by studying the effect of this oncogene on thyroid-hormone-responsive genes (Sap *et al.* 1989). As expected, the v-*erbA* gene product was able to abolish the responsiveness of such genes to thyroid hormone by binding to the thyroid hormone-response elements in their promoters and preventing binding of the cellular ErbA protein–thyroid hormone complex (Fig. 8.17).

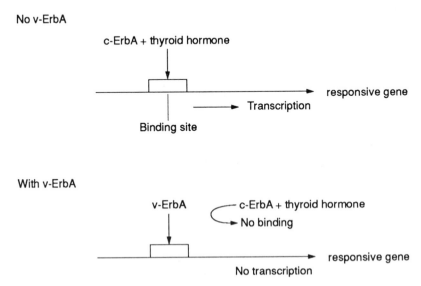

Figure 8.17 Inhibitory effect of the viral ErbA protein on gene activation by the cellular protein, in response to thyroid hormone. Note the similarity to the action of the α-two form of the c-ErbA protein, illustrated in Figure 7.31.

The explanation of how such gene repression by viral ErbA can result in transformation is provided by the observations of Zenke *et al.* (1988) who showed that the introduction of the viral gene into cells can repress transcription of the avian erythocyte anion transporter gene. This gene is one of those which is switched on when chicken erythroblasts differentiate into erythocytes. It has been known for some time that the viral ErbA protein can block this process, and it is now clear that this is achieved by blocking the induction of the genes needed for this to occur. In turn, such blockage of differentiation allows the cells to continue to proliferate. When this is combined with the introduction of the v-*erbB* gene, which encodes a truncated form of the epidermal growth factor receptor (Downward *et al.* 1984) and renders cell growth independent of external growth factors, transformation results (Fig. 8.18).

Transformation caused by v-*erbA* thus represents an example of the activation of an oncogene by mutation resulting, in this case, in its ability to act as a dominant repressor of transcription. As discussed in

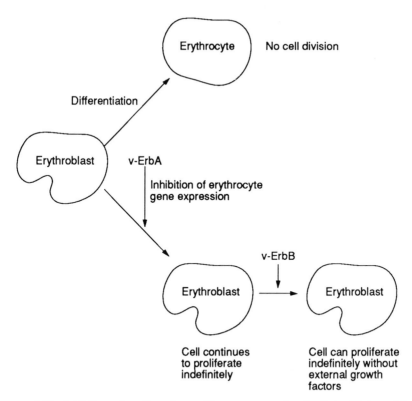

Figure 8.18 Inhibition of erythrocyte-specific gene expression by the v-ErbA protein prevents erythrocyte differentiation and allows transformation by the v-ErbB protein.

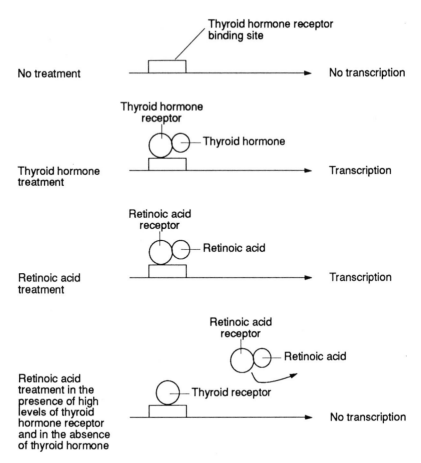

Figure 8.19 The thyroid hormone receptor can bind to its DNA binding site in the absence of thyroid hormone and prevent binding of the retinoic acid–retinoic acid receptor complex to the identical binding site, thereby preventing gene activation in response to retinoic acid.

Chapter 7 (Section 7.4.2), however, one alternatively spliced transcript of the c-*erbA* gene is also able to do this and, like the v-*erbA* gene product, cannot bind thyroid hormone. Hence this repression of transcription by a non-hormone-binding form of the receptor is likely to be of importance in normal cells also.

Indeed, such repression may be one facet of a complex network of regulatory controls involving the thyroid hormone-receptor in intact cells. Thus the DNA sequence element mediating response to retinoic acid is identical to that which renders a gene inducible by thyroid hormone (Table 6.3). In agreement with this, retinoic acid bound to its cellular receptor can activate transcription from thyroid-hormone-responsive genes. In the absence of thyroid hormone, however, the thyroid hormone-receptor binds to this site and blocks the binding of the retinoic acid complexed with its receptor (Graupner *et al.* 1989). Since the thyroid hormone-receptor will not activate transcription of the gene in the absence of thyroid hormone, this has the effect of preventing transcription (Fig. 8.19). This inhibition of retinoic acid-induced transcription by an unoccupied thyroid hormone-receptor is clearly similar to the blockage of thyroid-hormone-induced tran-scription by a form of the receptor that cannot bind hormone. Hence the dominant repression of transcription produced by the v-*erbA* gene product has opened up a new area involving the interaction of different receptors and different forms of the same receptor in the regulation of gene expression.

8.4.3 Other transcription-factor-related oncogenes

Although the *fos/jun* and *erbA* cases represent the best examples of the connection between oncogenes and transcription factors, several other cellular oncogenes encode nuclear proteins which are likely to be transcription factors (see Table 8.1).

One of these, the Myc protein, has been studied intensively in view of its overexpression in many human tumours (see Section 8.3) and there is considerable evidence suggesting that it does indeed encode a cellular transcription factor (reviewed by Cole, 1986). Thus the c-*myc* gene product can increase transcription of the 70 kDa heat-shock protein gene (hsp70) (Kingston *et al.* 1984) and the protein is known to contain the leucine zipper motif characteristic of many transcription factors, including Fos and Jun (see Section 7.2.4), as well as a helix-loop-helix motif (see Section 7.2.5). Moreover, mutations in the leucine zipper region of the protein abolish its oncogenic ability to transform normal cells (Dang *et al.* 1989), suggesting that the ability to act as a transcription factor is essential for Myc-induced transformation.

Despite all this evidence, the actual role of Myc in transcriptional control remains unclear. This is because it has not yet proved possible to demonstrate the binding of Myc to a specific DNA sequence in the manner that has been shown to occur for Jun and ErbA. It seems likely that Myc may resemble Fos in requiring another factor for such sequence-specific binding, and that once this factor has been identified the role of Myc will become clearer.

Alternatively, it is possible that the Myc protein may activate gene expression at a level other than transcription. In agreement with this idea, Prendergast & Cole (1989) identified two cellular genes whose expression was increased by entry of the Myc protein into cells, without any change in their transcription. In these experiments the

Myc protein appeared to be enhancing gene expression at a post-transcriptional level, possibly involving enhanced RNA splicing or transport of the spliced RNA from nucleus to cytoplasm.

In contrast to the case of Myc, sequence-specific DNA-binding activity has been demonstrated for the *myb* oncogene product (Biedenkapp *et al*. 1988) which binds to the sequence: pyrimidine–AACG/TG. Hence, like the other cases discussed in this section, this protein also is likely to play an important role in transcriptional regulation in both normal and cancer cells.

Although our discussion of oncogenes as transcription factors has focused on the cellular genes which have been picked up by RNA tumour viruses, it is worth noting that DNA viruses that can cause cancer also encode oncogenes capable of regulating cellular gene expression (reviewed by Kingston *et al*. 1985). In particular, both the large T oncogenes of the small DNA viruses, SV40 and polyoma, and the E1a protein of adenovirus are capable of affecting the transcription of specific cellular genes, and this ability is critical for the ability of these proteins to transform cells. Unlike the oncogenes of RNA viruses, however, the genes encoding these viral proteins do not appear to have specific cellular equivalents, and are likely to have evolved within the virus rather than having been picked up from the cellular genome.

The similar ability of oncogenes from DNA and RNA tumour viruses to affect cellular gene expression, despite their very different origins, suggests that such modulation of cellular gene expression is critical for the transforming ability of these viruses.

Interestingly, the anti-oncogene *Rb-1* (see Section 8.2) contains zinc finger motifs (see Section 7.2.3) and can bind to DNA (Lee *et al*. 1987), suggesting that genes of this type may also encode transcription factors capable of affecting the expression of cellular genes.

8.5   CONCLUSIONS

The study of the processes whereby increased expression or mutation of certain cellular genes can cause cancer has increased greatly our knowledge of the process of transformation, whereby normal cells become cancerous. Similarly, the recognition that the products of these cellular genes play a critical role in the growth regulation of normal cells has allowed insights obtained from studies of their activities in tumour cells to be applied to the study of normal cellular growth control.

Such a reciprocal exchange of information is well illustrated in the case of the oncogenes that encode cellular transcription factors. Thus, the isolation of the *fos* and *jun* genes in tumorigenic retroviruses has aided the study of the effects of growth factors on normal cells, while the recognition that the v-*erbA* gene product is a truncated form of the thyroid hormone-receptor has allowed the elucidation of its role in transformation via the inhibition of erythroid differentiation.

These two cases also illustrate the two mechanisms by which cellular genes can become oncogenic. Thus in the case of the v-*erbA* gene, mutations have rendered the protein different from the corresponding c-*erbA* gene from which it was derived. These alter the properties of the protein so it cannot bind thyroid hormone, and it behaves as a dominant repressor of transcription. Similar activation by mutation is also seen in oncogenes where products function at other stages of the growth-signalling process. Notable examples include the truncation of the epidermal growth factor receptor in the v-*erbB* oncogene, and the loss of GTP hydrolysing ability in the *ras* oncogene, both of which result in proteins 'frozen' into a positive growth-signalling form.

In contrast, oncogenes whose products are made only in response to a particular growth signal and whose activity then mediates cellular growth can cause cancer simply by the normal product being made at an inappropriate time. Thus, in the case of Fos or Jun, which are synthesized in normal cells in response to treatment with growth-promoting phorbol esters or growth factors, their continuous synthesis directed by a retroviral promoter is sufficient to transform the cell. A similar case is that of the *sis* oncogene, which encodes the cellular gene for platelet-derived growth factor and also causes cancer when overexpressed by a retrovirus.

Such cases of high-level expression of a normal oncogene product causing cancer also occur in a number of cases without any evidence of retroviral involvement. These examples of alterations in cellular regulatory processes producing increased expression of particular genes obviously provide another aspect to the connection between cancer and gene regulation. Thus, information on the origin of the translocations of the c-*myc* gene in Burkitt's lymphoma is obviously important for the study of cancer aetiology. Similarly, the fact that such translocations increase expression in some cases by removing elements that normally inhibit c-*myc* expression, will allow the characterization of such negative elements and their role in regulating c-*myc* expression in normal cells.

Hence the study of cellular oncogenes has contributed greatly to our knowledge both of cancer and of cellular growth-regulatory processes, and is likely to continue to do so in the future.

REFERENCES

Angel, I. P., M. Imagawa, R. Chiu, B. Stein, R. J. Imbra, J. J. Rahmsdorf, C. Jonat, P. Herlich & M. Karu 1987. Phorbol ester-inducible genes contain a common *cis* element recognized by a TPA-modulated *trans*-acting factor. *Cell* **49**, 729–39.

Angel, P., E. A. Allegretto, S. T. Okino, K. Hattori, W. J. Boyle, T. Hunter & M. Karn 1988. Oncogene *jun* encodes a sequence-specific *trans*-activator similar to AP-1. *Nature* **332**, 166–70.

Beug, H., P. Kahn, B. Vennstrom, M. J. Hayman & T. Graf 1985. How do retroviral oncogenes induce transformation in mammalian cells? *Proceedings of the Royal Society of London, Series B* **226**, 121–6.

Biedenkapp, H., U. Borgmeyer, A. E. Sippel & K.-H. Klempnauer 1989. Viral *myb* encodes a sequence-specific DNA-binding activity. *Nature* **335**, 835–7.

Bishop, J. M. 1987. The molecular genetics of cancer. *Science* **235**, 305–11.

Bos, T. J., D. Bohmann, H. Tsuchie, R. Tjian & P. K. Vogt 1988. V-*jun* encodes a nuclear protein with enhancer binding properties of AP-1. *Cell* **52**, 705–12.

Cole, M. D. 1986. The *myc* oncogene: its role in transformation and differentiation. *Annual Review of Genetics* **20**, 361–84.

Collins, S. & M. Groudine 1982. Amplification of endogenous *myc*-related DNA sequences in a human myeloid leukemia cell line. *Nature* **298**, 679–81.

Cooper, J. A. & P. Whyte 1989. R.B. and the cell cycle: Entrance or exit? *Cell* **58**, 1009–11.

Cory, S. 1986. Activation of cellular oncogenes in hemopoietic cells by chromosome translocation. *Advances in Cancer Research* **47**, 189–234.

Curran, T. & B. R. Franza 1988. Fos and Jun: the AP-1 connection. *Cell* **55**, 395–7.

Dang, C. V., M. McGuire, M. Buckaire & W. M. F. Lee 1989. Involvement of the leucine zipper region in the oligomerization and transforming activity of human c-*myc* protein. *Nature* **337**, 664–6.

De Feo-Jones, D., E. M. Scolnick, R. Koller & R. Dhar 1983. *Ras*-related sequences identified and isolated from the yeast *Saccharomyces cerevisciae*. *Nature* **306**, 707–9.

Downward, J., Y. Yarden, E. Mayes, G. Scrace, N. Totty, P. Stockwell, A. Ullrich, J. Schlessinger & M. D. Waterfield 1984. Close similarity of epidermal growth factor receptor and v-*erb-B* oncogene protein sequences. *Nature* **307**, 521–7.

Grandchamp, B., C. Picat, V. Mignotte, J. H. P. Wilson, K. TeVelde, L. Sandkuyl, P. H. Romeo, M. Goossens & Y. Nordmann 1989. Tissue-specific splicing mutation in acute intermittent porphyria. *Proceedings of the National Academy of Sciences of the USA* **86**, 661–4.

Graupner, G., K. N. Wills, M. Tzukerman, X.-K. Zhang & M. Pfuli 1989. Dual regulatory role for thyroid-hormone receptors allows control of retinoic acid receptor activity. *Nature* **340**, 653–6.

Halazonetis, T. D., K. Georgopoulos, M. E. Greenberg & P. Leder 1988. C-*jun* dimerizes with itself and with c-*fos* forming complexes of different DNA binding abilities. *Cell* **55**, 917–24.

Hayward, W. S., B. G. Neel & S. M. Astin 1981. Activation of a cellular onc gene by promoter insertion in ALV-induced lymphoid leukosis. *Nature* **290**, 475–80.

Kingston, R. E., A. S. Baldwin & P. A. Sharp 1984. Regulation of heat shock protein 70 gene expression by c-*myc*. *Nature* **312**, 280–2.

Kingston, R. E., A. S. Baldwin & P. A. Sharp 1985. Transcription control by oncogenes. *Cell* **41**, 3–5.

Lamph, W. W., P. Wamsley, P. Sassono-Corsi & I. M. C. Verma 1988. Induction of proto-oncogene Jun/Ap-1 by serum and TPA. *Nature* **334**, 629–31.

Leder, P., J. Battey, G. Lenoir, C. Moulding, M. Murphy, H. Potter, T. Stewart & R. Taub 1983. Translocations among antibody genes in human cancer. *Science* **222**, 765–71.

Lee, W., P. Mitchell & R. Tjian 1987. Purified transcription factor AP-1 interacts with TPA-inducible enhancer elements. *Cell* **49**, 741–52.

Lee, W.-H., J.-Y. Shew, F. D. Hong, T. W. Sery, L. A. Donso, R. Bookstein & E. Y.-H. P. Lee 1987. The retinoblastoma susceptibility gene encodes a nuclear phosphoprotein associated with DNA binding activity. *Nature* **329**, 642–95.

McGrath, J. P., D. J. Capon, D. V. Goeddel & A. D. Levinson 1984. Comparative biochemical properties of normal and activated human ras p21 protein. *Nature* **310**, 644–9.

Marth, J. D., R. W. Overell, K. E. Meier, E. G. Krebs & R. M. Perlmutter 1988. Translational activation of the *lck* proto-oncogene. *Nature* **332**, 171–3.

Perry, R. P. 1983. Consequences of *myc* invasion of immunoglobulin loci, facts and speculation. *Cell* **33**, 647–9.

Prendergast, G. C. & M. D. Cole 1989. Post-transcriptional regulation of cellular gene expression by the c-*myc* oncogene. *Molecular and Cellular Biology* **9**, 129–34.

Rabbits, T. H. 1985. The c-*myc* proto-oncogene: involvement in chromosomal abnormalities. *Trends in Genetics* **2**, 327–31.

Rauscher, F. J., D. R. Cohen, T. Curran, T. J. Bos, P. K. Vogt, D. Bohmann, R. Tjian & B. R. Franza 1988. Fos-associated protein P39 is the product of the c-*jun* oncogene. *Science* **240**, 1010–16.

Reith, W., S. Satola, C. H. Sanchey, I. Amaldi, B. Lisowska-Grospiere, C. Griscelli, M. R. Hadam & B. Much 1988. Congenital immunodeficiency with a regulatory defect in MHC Class II gene expression lacks a specific HLA-DR promoter binding protein RF-X. *Cell* **53**, 897–906.

Sap, J., A. Munoz, A. Schmitt, H. Stunnenberg & A. Vennstrom 1989. Repression of transcription at a thyroid hormone response element by the v-*erbA* oncogene product. *Nature* **340**, 242–4.

Sap, J., A. Munoz, K. Damm, Y. Goldberg, J. Ghysdael, A. Leutz, H. Berg & B. Vennstrom 1986. The c-*erb-A* protein is a high-affinity receptor for thyroid hormone. *Nature* **324**, 635–40.

Schuermann, M., M. Neuberg, J. B. Hunter, T. Jenuwein, R.-P. Ryseck, R. Bravo & R. Muller 1989. The leucine repeat motif in *fos* protein mediates complex formation with Jun/Ap1 and is required for transformation. *Cell* **56**, 507–16.

Takeya, T. & H. Hanafusa 1985. Structure and sequence of the cellular gene homologous to the RSV-*src* gene and the mechanism for generating transforming virus. *Cell* **32**, 881–90.

Vogt, P. K., T. J. Bos & R. F. Doolittle 1987. Homology between the DNA-binding domain of the GCN4 regulator protein of yeast and the carboxyl-terminal region of a protein coded for by the oncogene *jun*. *Proceedings of the National Academy of Sciences of the USA* **84**, 3316–19.

Waterfield, M. D., G. J. Scrace, N. Whittle, P. Stroobant, A. Johnson, A. Wastesan, B. Westermark, C.-H. Heldu, J. S. Huang & T. F. Deuel 1983.

Platelet-derived growth factor is structurally related to the putative transforming protein p28 *sis* of simian sarcoma virus. *Nature* **304**, 35–9.

Weinberg, R. A. 1985. The action of oncogenes in the cytoplasm and nucleus. *Science* **230**, 770–6.

Weinberger, C., C. C. Thompson, E. S. Ong, R. Lebo, D. J. Gruol & R. M. Evans 1986. The c-*erb-A* gene encodes a thyroid horome receptor. *Nature* **324**, 641–6.

Zenke, M., D. Kahn, C. Disela, B. Bennstrom, A. Leutz, K. Keegan, M. J. Hayman, H.-R. Chui, N. Yew, J. D. Engel & H. Berg 1988. V-*erb-A* specifically suppresses transcription of the avian erythrocyte anion transporter (Band 3) gene. *Cell* **52**, 107–19.

CHAPTER NINE

Conclusions and future prospects

The extraordinary rate of progress in the study of gene regulation can be gauged from the fact that only 10 years ago, no transcriptional regulatory protein or its DNA-binding site in a regulated promoter had been defined. The tremendous advances since that time should be evident from this book. For example, in the case of gene induction by steroid hormones we now know that these agents act by activating specific receptor proteins (Ch. 7, Section 7.4.3), that such activated receptors bind to specific DNA sequences upstream of the target gene (Ch. 6, Section 6.2.3), displacing a nucleosome (Ch. 5, Section 5.6.2), and that a region in the receptor protein then interacts with a factor bound to the TATA box to activate transcription (Ch. 7, Section 7.3.3).

Clearly, this process is reasonably well understood in outline. What remains to be obtained is the detailed mechanism of each of these stages. How, for example, does binding of the hormone promote dissociation from hsp90? How does the activation domain of the receptor interact with the TATA box binding factor, and how does this interaction facilitate transcription? The answers to these questions will require a move from the DNA encoding the receptor to an understanding of its actual protein structure and that of the other proteins with which it interacts. This will pave the way for a structural analysis of mutant receptors defective in each of the activities described above and an eventual understanding of the inter-molecular interactions, which are central to transcriptional regulation.

Although regulation by steroid hormones or other transcriptional inducers, such as cyclic AMP or phorbol esters, represent the best-understood systems, our ultimate aim must be an understanding of the regulation of tissue-specific gene expression explaining, for example, why the albumin gene is expressed only in the liver or the myosin gene only in muscle.

Fortunately, the available evidence suggests that similar regulatory processes are involved in these systems also. To consider one example only, a regulatory protein (MyoD) involved in skeletal muscle cell differentiation has been identified. Introduction of this gene into a mouse fibroblast line is sufficient to cause it to differentiate into skeletal muscle cells (Davis *et al.* 1987) and the activation of this gene is likely to be the cause of the production of this phenotype when these cells are treated with the demethylating agent, 5-azacytidine (see Ch. 5, Section 5.5.1).

The MyoD protein is a sequence-specific DNA-binding protein, which has been shown to bind to elements in the promoters of muscle-specific genes and activate their transcription, thereby producing muscle cells. It contains a helix-loop-helix motif and a basic DNA-binding region similar to those found in other transcriptional activators (see Ch. 7, Sections 7.2.4 & 7.2.5) and in the *myc* oncogene product (Ch. 8, Section 8.4.3). These elements are necessary for the muscle-inducing activity of MyoD (Tapscott *et al.* 1988). It is clear, therefore, that this gene is involved in inducing muscle development and that it does so by binding to DNA and affecting the transcription of other genes, using protein domains found also in the proteins modulating gene expression in response to chemical inducers.

It is likely, therefore, that further such studies will allow a detailed understanding of the processes regulating tissue-specific gene expression. Eventually, however, it will be necessary to explain not only how liver-specific genes are expressed only in the liver but also why liver cells are produced only at one place in the body, brain cells at another, and so on. Such studies on the development of the pattern of the body are most advanced in *Drosophila* because of the availability of mutations in homeotic and other genes which control these processes (Ch. 7, Section 7.2.2).

A particularly interesting example of how such proteins may regulate spatial differences in gene activity is provided by the product of one such gene, the *bicoid* gene. This protein, which is vital for the development of the anterior half of the fly, is found at highest concentration at the anterior tip of the developing embryo, with the protein concentration declining progressively posteriorly. Recent studies (Driever *et al.* 1989) have shown that genes activated by the *bicoid* gene product contain binding sites in their promoters which have either high affinity or low affinity for the bicoid protein. Genes with low-affinity binding sites are only activated at high concentration of the protein and hence will be expressed only at the extreme anterior end of the developing fly. In contrast, genes with high-affinity binding sites are active at much lower protein concentrations and will be active both at the anterior end and more posteriorly.

Hence, a single protein can produce spatial variation in gene activity by the simple means of a concentration gradient which is first established in the unfertilized egg. When this is coupled with the possibility of interactions between different proteins, with some genes being active when both proteins are present, some in the presence of only one, and so on, it becomes possible to see how the development of specific organs in specific places can be achieved through specific proteins regulating gene activity. Clearly, much more work remains to be done before this process is understood fully. The progress in the past 10 years suggests, however, that this is not impossible and that an understanding of *Drosophila* and eventually mammalian development in terms of differential gene expression can be achieved.

REFERENCES

Davis, H. L., H. Weintraub & A. B. Lassar 1987. Expression of a single transfected cDNA converts fibroblasts to myoblasts. *Cell* **51**, 987–1000.

Driever, W., G. Thoma & C. Nusslein-Volhard 1989. Determination of spatial domains of zygotic gene expression in the *Drosophila* embryo by the affinity of binding sites for the bicoid morphogen. *Nature* **340**, 363–7.

Tapscott, S. J., R. L. Davis, M. J. Thayer, P.-F. Cheng, H. Weintraub & A. B. Lassar 1988. MyoD1: a nuclear phosphoprotein requiring a *myc* homology region to convert fibroblast to myoblast. *Science* **242**, 405–11.

Index

acetylcholine receptor 129
acid blobs 211
actin 10, 68
activation domains 153, 208–15
   acidic 211–12, 214, 223
   glutamine-rich region 212, 214, 226
   proline-rich regions 212, 226
activator molecules 102
acute intermittent porphyria 231
adenovirus 78, 93
   transactivator, E1a 200, 252
   translation of viral mRNAs 91
albumin 58, 144, 155, 257
   promoter 151
alcohol dehydrogenase 72
aldolase A 72
Alu repeated sequence 164, 167, 174
α-amanitin 138
Amphibia
   lampbrush chromosomes 61–2
   nuclear transplantation 21–2
   regeneration 19
   transdifferentiation 19
   U1 RNA 80
amphipathic helix 211–13, 215, 226
amplification of genes 13, 24–30, 43–4, 243
   see also DNA amplification
α-amylase 72
   gene 71, 155
antibodies 4–6, 37–43, 73, 131, 157–8
   production 37–43, 81, 140
antigenic variation 37–8
anti-oncogenes 237
   Rb-1 237, 252
α-1-antitrypsin 149, 155, 219
apolipoprotein B 81–2
   apo-B48 82
   apo-B100 82
arabinose operon 101
auto-regulation 87
avian erythroblastosis virus (AEV) 246
avian erythrocyte anion transporter gene 249
avian leukosis virus (ALV) 237–40
avian sarcoma virus ASV17 243
5-azacytidine 124, 258

bacteriophage 434 repressor protein 189, 191

B-cells 13, 19, 72–3, 75, 81, 84, 103, 140, 151, 155, 158–9, 182, 217–18
   antibody production 37–43, 141
   transformation of 239–40
Bombyx mori, see silk moth
brain tissue 9, 16, 23, 58–9, 77, 80, 103, 117, 120, 166–7, 194
Britten and Davidson model 141–2, 148, 150, 157, 165, 168, 170, 217
Burkitt's lymphoma 239, 253

calcitonin 77–8, 80
calcitonin/CGRP,
   mRNA 80
   polyadenylation sites 76, 78
   RNA splicing factors 78, 80
calcitonin-gene-related peptide (CGRP) 76–8, 80–1
   alternative splicing of transcripts 76–8
   mechanism 78, 80
cancer 231–53
carrot, regeneration 19–20
cartilage-producing cells 103–4, 115
casein kinase 6
casein mRNA, stability 83–5, 90
   cassette mechanism 33
CCAAT box 143–4, 212
cell lineage 35–6, 104, 115–16, 121, 123, 180
chloroplast mRNAs, stability 85
chondroitin sulphate 104
chorion genes, amplification 28–30, 44
chromatin
   beads on a string structure 109–10, 112, 116–17, 134–5
   DNaseI hypersensitive sites 126–34, 147, 151, 159, 198
   sensitivity to DNaseI digestion 113–16, 118, 120, 123–4, 126–7, 134
   mechanism 116–18
   structure 30, 106–18, 138, 170, 180
   changes of 112–18, 152–3, 159
   open 114–17, 120, 123, 168
   regulation 167–8
   solenoid 109, 111, 116–17, 126, 134–5
   supercoiling 131
   undermethylation 119–20, 123, 126–7, 134
chromosomes 14–15, 17–18, 109
   double minute 243

lampbrush 61–2
loss of 17–19
polytene 25–6, 59–61, 131, 143
puffs 59–61
translocations 239–40, 253
chymotrypsin gene 156
collagen 11, 129
collagenase 144, 149
α2 (I) collagen gene 125
conalbumin 149
congenital severe combined
    immunodeficiency 231
consensus sequences 144–50, 159
copper, binding to ACE1 protein 221–2
crustaceans, chromosome loss 17
crystallins 19
cyclic AMP 150, 216, 224–5, 227, 257
response element (CRE) 144, 150, 224
cysteine protease 129

Davidson and Britten model 54–6, 59, 169
deoxyribonuclease I (DNaseI) 113–18, 120,
    123–4, 126–7, 130, 173
hypersensitive sites 126–34, 147, 151,
    159, 198
Dictyostelium discoideum 58
differentiated state
commitment to 103–6, 115, 120–1, 123,
    134, 187–8
stability of 20, 105
dihydrofolate reductase 129, 144
gene amplification 29
dimethyl sulphoxide 49, 53, 115
DNA
amplification 13, 24–30, 43–4, 243
    chromosomal studies 25–8
    molecular studies 28–30
binding by transcription factors 185–208,
    225
circular 132
complementary (cDNA) 7–8
deletion events 123
demethylation 121–5
linker 109
looping out of 160–1
loss 13–24, 43–4, 106
    chromosomal studies 14–19
    functional studies 19–22
    molecular studies 22–4
macronuclear 25–7
methylation 118–26
    hemi-methylated sites 121–2
    tissue-specific pattern 119–20
rearrangement 13, 30–44
    immunoglobulin genes 37–43
    molecular studies 31–2
    trypanosomal 37
    yeast mating type 32–7

repeated 164–5
5S 195–6
satellite 17
sequence elements 139, 151, 153
sequencing 24, 32, 43
single-stranded 131
unmethylated 121–4
viruses 252
    adenovirus 78, 91, 200, 252
    polyoma 252
    SV40 78, 131–2, 159, 161, 252
    see also Z-DNA
DNA-binding
domain 201, 205–7, 209–11, 220, 243,
    258
proteins 151, 189, 191–2, 199, 221, 244,
    247, 258
    sequence-specific 191, 194, 196, 200,
    203
DNase I, *see* deoxyribonuclease I
domain-swap experiments 208–10
Drosophila melanogaster
chorion gene amplification 28–30, 44
chromosomes
    polytenization 25–6, 59–60, 131, 143
    puffs 59–61, 131, 143
commitment to differentiated state
    104–6, 115, 187–8
development 70, 258–9
DNA methylation 124
genes
    bicoid 258
    daughterless 207, 226
    fushi tarazu (ftz) 187–8
    homeotic 186–7, 226, 258
    ultrabithorax 81, 188
heat shock 60
    genes 128, 144, 146, 159
    proteins 59, 61
histones 125
homeobox proteins 37, 192, 194
homeodomains 190
mutants 185–92
oncogene equivalents in 236
P-element transposase 69–70
processing versus discard decisions 68–9
proteins
    Glass 197
    glue 129–30
    Hunchback 197
    Kruppel 196–7, 226
    Snail 197
    yolk 58
sex determination 70, 81
suffix element 164
transcription factors 212–13
Zeste transcriptional regulator 212
dyad symmetry 148–50

ecdysone 59–61
elastase 129
  gene 155–6
enhancers 42, 130–2, 153–64, 170, 180,
    238–40
  heat-inducible 159
  mechanism of action 158–61
  tissue-specific activity 155–8, 162–3
enhansons 159
epidermal growth factor 65
  receptor 253
    mRNA stability 84
erythroblasts 16–17, 49, 249
erythocytes 16–17, 57, 115, 117, 119–20,
    127, 130, 249
  loss of nucleus 16–17, 44
Escherichia coli
  arabinose operon 101
  lac operon 140
  Sp1 protein 196
exons 63–4
exonuclease III 146–7
eye tissues 19

ferritin RNA
  stem-loop structures 86, 93–5
  translational control of 91, 95
fertilization 91, 96
fibroblasts 49, 73, 84, 104, 124, 155
fibroin gene 28, 129
fibronectin 73, 144, 149
finger-swop experiments 200–1
α-foetoprotein
  gene 128–9, 155
  mRNA 10, 23, 51, 58
footprint assay 173
Friend erythroleukaemia cells 49, 53, 84, 115
fusion proteins 182

GAL-4, *see* yeast transcription factors
β-galactosidase 46, 139, 182, 213
G-banding 14–16, 19
gel electrophoresis 2–4
genes
  a 33–7
  α 33–7
  bicoid 258
  daughterless 207, 226
  engrailed 187
  era-1 220
  ftz 187–8
  heat-shock 128, 134, 143–50, 159, 251
  HLA class II 231
  HO 34–5
  homeotic 186–8, 226, 258
  housekeeping 143
  integrator 142, 148, 217
  lck 242

myc 239–40
sqs4 130
silent 36, 38, 126
src 233
ubx 81, 188
unc-86 194, 226
VSG 37–8
see also oncogenes, proto-oncogenes,
    named proteins
germ cells 17, 69–70
Giemsa stain 14–15, 19
globin 10, 16, 49, 53, 93, 129, 140
  α-globin 140–1, 144
    gene 141
  β-globin 23, 57, 123, 140–1, 144
    gene 132, 141, 153, 155
  γ-globin 123, 141
  ε-globin 167
  gene 53–4, 114–15, 117, 119–20, 127
  RNA 49, 53–4
  translational control of 91, 93
α-2 globulin gene 151
glucocorticoid hormones 101
  receptor 150–3, 198, 200–3, 209–11, 213,
    222–3
  /steroid complex 134
  responsive genes 134, 150–1, 200
glucose-6-phosphate dehydrogenase 129
glycinin 58
α-gonadotrophin 144, 149
growth hormone gene 155, 246

heart tissue 80–1
heat shock 93, 96, 100
  consensus element 144–50, 159
  genes 128, 134, 143–50, 159, 251
    hsp27 144, 149
    hsp70 143–50, 159, 251
    hsp83 144, 149
  mRNAs
    stability of 97
    5' untranslated region 93
  proteins 59, 61, 91, 93, 96–7, 129, 143,
      222–3, 225, 227, 251, 253
    hsp 70 61, 143, 251
    hsp 90 222–3, 225, 227, 257
  response element 143–8
  transcription factor (HSTF) 133–4, 147–9,
      151
HeLa cells 84, 124, 217–18
helix-loop-helix motif 207, 226, 251, 258
helix-turn-helix motif 185–94, 196, 198,
    203–4, 207
hemi-methylated sites 121–2
hepatocytes 73
herpes simplex virus 91, 144–5
  trans-activating protein VP16 210–11, 223
histones 107, 129, 132, 134

acetylation 126
H1 107, 109–10, 116–17, 120
H2A 107, 125–6, 153
H2B 107, 192
H3 85, 107, 126
H4 107, 126
modifications 125–6
mRNA
  degradation signals 89
  stability 84–5, 88, 90
octamer 107, 109
proteins 107
ubiquitination 125–6
homeobox 185–92, 197, 221, 226
  proteins 37, 191–2, 194, 197
homeobox-like sequence 194
homeodomain 188–90, 193–4
homeotic genes 186–8, 226, 258
  engrailed 187
  fushi tarazu 187–8
  ultrabithorax (*ubx*) 81, 188
homothallism gene (*HO*) 34–5
hormone
  binding proteins 247–8
  receptors 135, 221–2, 246–8
    see also named hormones
  responsive genes 134, 150–1, 200, 222–3
human immunodeficiency virus, Rev
    protein 83
hybridization 7–8
  in situ 10–11

Ia antigen-associated invariant chain 94, 96
identifier repeat (ID) 166–7, 170
imaginal discs 104–6, 115, 187–8
immunodeficiency 231
immunofluorescence 5, 158
immunoglobulin 13–14, 38, 129
  constant (C) region 39–41, 240
    genes 40–3, 155
  enhancers 155, 159, 161, 240–1
  genes 39–40, 44, 140, 151, 157–8, 182,
    194, 217–18, 239–41
    diversity segment 41–3
    enhancer element 42–3, 155, 159, 161
    γ light chain 40–1
    κ light chain 41
    point mutations 42
  heavy chains 38–9, 41–2, 72–3, 140–1
    alternative splicing of transcript 75–7,
      81
    enhancers 155, 159, 161
    genes 140, 151, 239
    polyadenylation sites 76
  joining (J) region 240
    genes 40–3, 155
  light chains 38–41, 140–1
    enhancers 155

genes 140, 151
  membrane-bound 73, 75, 81
  promoters 159, 240
  secreted 75, 81
  variable (V) region 39–41
    genes 40–3
initiation
  codons 94, 96
  factor eIF2 93
insects, chromosome loss 17
insulin
  gene 155–8
  production 58
integrator genes 141–2, 148, 217
β-interferon 129, 149
  enhancer element 163
  mRNA, regulation of translation 96
internal control regions 171–6, 194
intervening sequences 46, 48–9, 52, 63, 68,
    70, 164
intestinal tissues 82
introns 46, 48–9, 68, 70
iris epithelium 19
iron response elements 93
isoelectric focusing 3

kidney tissue 9, 58–9, 103, 167

lac operon 100, 120, 139–40
lens regeneration 19–20, 105
leucine zipper 204–7, 226, 244, 246, 251
liver tissue 6, 9, 16, 23, 49, 51, 58–9, 67,
    71–2, 82, 84, 102–3, 113, 116, 119,
    128, 151, 155, 157–8, 167, 204, 219,
    257
loci
  HML 33–4, 36
  HMR 33–4, 36
  MAT 33–4
lysozyme 149

macronuclei 25, 27, 131
maintenance methylation 121–3
mammary gland tissue 6, 84, 90
messenger RNA, *see* mRNA
metallothionein II A 144, 149–50
  gene 221–2
  promoter 80
methotrexate 29
5-methyl cytosine 118, 120, 123–4
Miastor 17
micrococcal nuclease 109, 113, 130–1
micronuclei 25
mini-chromosome 132
monocytes 72
mRNA
  monocystronic 140
  polycistronic 139–40

regulated degradation 89
stability 83–5, 89–90, 95, 97, 241
stem-loop structures 85–6, 88
tissue-specific expression 6–12, 48
3' untranslated region 85
muscle 10, 13–14, 19, 49, 72, 81, 103, 124, 257–8
myelin basic protein 73
myosin 10, 13, 19, 257
  heavy chain 72, 149
    gene 246
  light chain 72
    alternative splicing 74
    gene 71

negative noodles 211
nematodes
  chromosome loss 17–18
  unc-86 gene 194, 226
neural tissue 73
Northern blotting 9–10, 23, 49–51, 167, 182
nuclear RNA 48–53, 57, 67, 83, 166–7
  degradation of 52–3, 56, 68
  transcription 138
nuclear run-on assays 56–9, 63
nuclear transplantation 21–2, 105, 123
nucleases
  Ba131 131
  S1 131
nucleosome 107–9, 112–14, 116, 125, 132–5, 147, 151–3, 159, 198, 257

octamer-binding proteins
  Oct-1 192, 194, 210–12, 226
  Oct-2 192, 212, 217–18, 223, 226
octamer motif 151, 159, 192
oestrogen 49–51, 57, 102, 115–16, 135
  receptor 149, 198, 200–1, 203, 210
  responsive element 149–50, 203, 210
(2'5')oligo A synthetase 72
oncogenes 233, 235–7, 253
  amplification 29, 243
  c-erbA 219, 247, 250, 253
  c-fos 89, 144, 149, 226, 244–5, 253
  c-jun 226, 244–5, 253
  c-myb 65, 252
  c-myc 63–4, 89, 129, 144, 163, 207, 226, 237–43, 251, 253, 258
    mRNA degradation 89
    mRNA stability 84, 90, 241
    promoter 240–1
    silencer element 163–4
  c-ras 129, 144, 253
  c-src 233–5, 242
  cellular 235
  elevated expression of 237–43
  L-myc 243
  large T 252

N-myc 243
sis 253
transcription factors as 243–52
v-erbA 246–51, 253
v-erbB 246, 249, 253
v-jun 243
viral 235
see also proto-oncogenes
oocytes
  lampbrush chromosomes 61–2
  Xenopus laevis 174–6
oogenesis 28
operon 139
ovalbumin 49, 129, 149
  gene 57, 102, 113–16, 127, 130
  mRNA 49–51
oviduct tissues 49–51, 57, 102, 113–16, 127, 130
Oxytrichia 25

pancreatic tissue 58, 155–8
Parascaris 18
P-element transposase 69–70
permease enzyme 139
phorbol esters 206, 218, 224, 245, 253, 257
phosphorylation 97, 142, 148, 221, 224–5, 227
pituitary gland 155, 194
plants, regeneration 19–20
plasmacytomas 239
polyadenylation 46, 73, 75–6, 96
polycistronic (multi-gene) mRNA 139–40
polytenization 25–7, 59–61
porphobillinogen deaminase 231
post-transcriptional regulation 48–9, 52, 55–6, 59, 67–97, 168–70, 227, 252
  of RNA splicing 68–82, 252
  of RNA stability 83–90
  of RNA transport 82–3, 252
post-translational modification 148
POU domain 192–4, 212, 226
pre-albumin gene 157–8
preproinsulin 129
preprotachykinin 73
pro-collagen, type I 144
  mRNA stability 84
progesterone-responsive genes 150
prolactin 129
  gene 155
  effect on casein mRNA stability 83, 85, 89
promoters 142–3, 147, 151, 153–5, 157–9, 161, 163, 180, 237–41
  c-myc 240–1
  immunoglobulin 240
  insertion 238, 240
  regulatory elements 151–3
  selection of 73

pro-myeloid cell line HL-60 63–4
protein kinase 224, 246
proteins
  ACE1 221–2
  ADR1 196–7
  Antennapedia 189–91, 212
  apo-B48 82
  apo-B100 82
  araC 101
  Bocoid 191, 258
  C/EBP 219
  Cro 189
  Cut 212
  E1a 200, 252
  engrailed (Eng) 187–8
  ErbA 246–9, 251
  Erb α-1 219–20
  Erb α-2 219–20, 248
  Fos 204–6, 243–6, 251, 253
  Ftz 187–8, 190
  functional domains 182
  fusion 182
  GAL4 200, 209, 211–13, 215, 222–4, 226
  GAL80 223–4
  GCN4 94–5, 97, 204–5, 208, 211, 221,
    226–7, 242–4
  heat-shock 59, 61, 91, 93, 96–7, 129, 143,
    222–3, 225, 227, 251, 253
  HMG 14 116–18, 120, 126
  HMG 17 116–118, 120, 126
  homeobox 37, 191–2, 194, 197
  housekeeping 4, 9, 13
  hsp70 61, 143, 251
  hsp90 222–3, 225, 227, 257
  Hunchback 197
  I ϰB 224
  Jun 204–6, 243–6, 251, 253
  Kruppel 196–7, 226
  Lck 242
  liver-specific 58, 67
  Myc 204, 251–2
  MyoD 258
  NF ϰB 182–3, 218, 224–5
  Oct-1 192, 194, 210–12, 226
  Oct-2 192, 212, 217–18, 223, 226
  p39 244
  Pit-1 194
  POU 194
  receptor 73, 101, 150–3, 198–204, 209–11,
    213, 222–3, 225, 227, 246–51, 253,
    257
  regulatory 34–5, 101, 176, 189, 192, 217
  repressor 100, 189, 191
  Rev 83
  Sgs4 130
  SIN3 34–5
  Sp1 196–7
  Src 233
  SW15 34–5
  TDF 197
  tissue-specific expression 1–12
  Ubx 188–90
  v-ErbA 249
  v-ErbB 249
  VP16 210–11, 223
  Xfin 196–7
  see also transcription factors, named
    enzymes and structural proteins
proto-oncogenes 231–7, 243
  c-*fos* 244
  lck 242
Protozoa, ciliated
  germ-line micronuclei 25
  polytenization 25–7
  rRNA amplification 27
  somatic macronuclei 25, 131
  see also trypanosomes
pulse-labelling studies 53–7, 59

receptor-hormone complex 134–5
receptor proteins 73, 101, 150–3, 198–204,
  209–11, 213, 222–3, 225, 227, 246–51,
  253, 257
recognition helix 191–2, 203
regeneration 19–20, 25
regulatory elements 141, 160, 170, 217
  common 141
  short 142–3, 151–3
regulatory processes
  long-term 103, 105–6
  short-term 103
regulatory proteins 34–5, 101, 176, 189,
  192, 217
repeated sequences
  role of 164–70
  short interspersed repeats (SINES) 164–7,
    174
  tissue-specific transcription 166–7, 170
restriction endonucleases 22, 31–2, 113,
  127
  *Bam*H1 40, 234
  *Eco*R1 23, 29
  *Hin*dIII 29
  *Hpa*II 118–20
  *Msp*I 118–20
  *Sac*I 234
reticulocytes 10, 16, 91, 93, 141
retinoblastoma 237
retinoic acid 198, 221, 250–1
retroviruses 232–3, 235, 237, 243–4, 246,
  253
ribosomal DNA, amplification 27–8
ribosomal RNA, *see* rRNA
RNA
  cap structure 46
  degradation 83

control of 89–90, 96
editing 81–2
maternal 91
monocystronic 140
populations of tissues 8–9
primary transcript 46, 49, 56, 63, 71, 73
    modification 46–7
splicing 40–1, 46, 49–51, 53, 63, 68–9, 82, 138
    alternative 69–78, 80–1, 96–7
    processing versus discard decisions 68–9, 83
    regulation of 68–82
stability
    changes 89–90
    regulation 83–90, 97
    3′ untranslated regions 85
transport 51, 53, 67
    regulation 82–3, 252
tumour viruses 252
see also mRNA, nuclear RNA, rRNA, transfer RNA, U RNA
RNA polymerase 46, 53, 61–4, 112, 132, 134, 138–9, 151, 185, 207, 214, 216–17
    I 138–9, 170, 172, 174
    II 138–9, 164–70, 172, 174, 196, 214
    III 138–9, 165–70, 172–6, 196, 217
Rôt curve 7–9, 52
    inter-tissue 8–9
Rous Sarcoma virus (RSV) 232–3
    src gene 233–4
rRNA 129, 138, 170
    amplification 27–8
    5S 129, 138–9, 171–4
        gene 171–6, 194, 217
    7SL 174

salivary gland tissues 71–2
SDS-polyacrylamide gel electrophoresis 2–3
sea urchin
    post-transcriptional regulation 59, 67
    nuclear RNA 52, 83
sensor elements 142
serum response factor 149, 207
silencer elements 163–4, 240–1
silk moth
    chorion genes 30
    fibroin 129
    silk glands 28–9
SINE (short interspersed repeats) elements 164–6
    Alu sequence 164, 167, 174
    B1 repeats 164
    B2 repeats 164, 166
    ID repeat 166
    suffix element 164

slime mould 58
solenoid structure 109, 111, 116–17, 126, 134–5
somatic cells 69
    chromosome loss 17–19
    mutation 42
somatostatin 144, 149–50
Southern blotting 22–3, 28–9, 31–2, 40, 43, 113–14, 127, 182, 233–4
soya bean
    embryo 51
    seed proteins 58
Sp1 box 144, 212, 226
Spisula solidissima 91–2
spleen tissue 23
spliceosome 80
splicing 41, 46, 49–51, 53, 63, 68–9, 82, 138, 219
    alternative 69–78, 96–7, 219, 231, 247, 250
    mechanism of 78, 80–1
    factors, tissue-specific 78, 80–1
    regulation 68–82, 219–21, 252
stem cells 16, 35–6, 123
stem-loop structures 85–6, 88, 93–5
steroid
    hormones 46, 59–60, 100–3, 116, 129, 134, 198, 257
    inducible genes 133
    receptor proteins 150, 199–200, 203–4, 225, 227
    regulated genes 198, 202
    responsive genes 101, 134–5, 203
steroid-thyroid hormone receptor gene super-family 198–9, 210, 219, 221, 226, 246–7
Stylonichia 131

tachykinins 81
TATA box 143–4, 147, 215–16, 257
    binding factor 134, 151–2, 215, 257
T-cells 43, 72, 218
    receptor 43
testis determining factor (TDF) 197
Tetrahymena 25
thymidine kinase 144
    gene 145
thyroid
    hormone 219–20, 247–8, 250–1, 253
        receptor 73, 198, 200, 203, 246–51, 253
        responsive element 150, 246, 248, 250
    tissue 73, 77
tissue plasminogen activator (t-PA) 96
tissues
    brain 9, 16, 23, 58–9, 77, 80, 103, 117, 120, 166–7, 194
    cartilage 103–4, 115
    embryonic yolk sac 10

erythroid 167
eye 19
foetal liver 23, 128
heart 80–1
intestinal 82
kidney 9, 58–9, 103, 167
liver 6, 9, 16, 23, 49, 51, 58–9, 67, 71–2,
    82, 84, 102–3, 113, 116, 119, 128,
    151, 155, 157–8, 167, 204, 219, 257
mammary gland 6, 84, 90
mRNA expression 6–12
muscle 10, 13–14, 19, 49, 72, 81, 103,
    124, 257
neural 73
oviduct 49–51, 57, 102, 113–16, 127, 130
pancreas 58, 155–8
pituitary gland 155
protein composition of 2–6, 12
RNA populations 8–9
salivary gland 71–2
spleen 23
tadpole intestinal epithelium 22
thyroid 77
tobacco plant
    nuclear RNA 52
    regeneration 19
trans-acetylase 139
transcription 46–65, 219
    activation of 102, 152, 160, 207–17, 225
    control of
        DNA sequence elements 138–76
        transcription factors 180–227
    initiation of 61, 63–4, 73
    regulation 48–61, 67, 73, 81, 90, 96–7,
        100–35, 219, 227
        at elongation 61–5, 242
        nuclear RNA studies 48–53
        nuclear run-on assays 56–9
        polytene chromosomes 59–61
        pulse-labelling studies 53–6
        RNA polymerases I and III 170–6
transcription complex 151, 172, 174, 207,
    214, 216–17
transcription factors 138, 141–2, 148, 151,
    153, 159, 167, 180–227, 231, 246,
    251–2
    activation 221–5
        protein-ligand interaction 221–3
        protein modification 224–5
        protein-protein interaction 222–3, 225
    ADR1 196
    AP1 206, 243–6
    AP2 150, 207, 212
    ATF 216–17
    basic DNA-binding domain 204–7
    C/EBP 204–5, 219, 226
    CREB 149, 224–5, 227
    CTF/NF1 207, 212, 226

DNA binding 185–207
    Drosophila 212–13
    heat-shock (HSTF) 133–4, 147–9, 151
    helix-loop-helix motif 207, 226, 251, 258
    helix-turn-helix motif 185–94, 196, 198,
        203–4, 207
        homeobox 185–92
    Jun 212
    leucine zipper 204–7, 226, 244, 246, 251
    mammalian 197, 213–16, 224, 244
    NGF-1A 197
    oncogenes 243–52
    proto-oncogenes 237
    Sp1 180–1, 196, 212, 214, 226
    TFIIC 216
    TFIID 215–17
    TFIIE 216
    TFIIIA 172–5, 194–7, 217, 226
    TFIIIB 172, 174
    TFIIIC 172, 174
    yeast 200, 208, 211–15, 221–4, 243
        see also yeast
    zinc finger motif 194–203, 226, 252
transdifferentiation 19, 123
transferrin 58
    receptor
        mRNA stability 85, 95
        mRNA stem-loop structures 86, 93–4
transfer RNA 129, 138, 172
transgenic mouse 157–8
translation 219, 241
    alternative 96
    regulation of 90–7, 221, 242
        3' untranslated region 94, 96
        5' untranslated region 94–6, 242
    start sites 242
translational control 90–1, 96
    mechanism of 91–6
transposable P element 69–70
transthyretin 219
trompomyosin 72
troponin T 72
    RNA splicing of 78–9, 81
trypanosomes
    antigenic variation in 37–8
    variant surface glycoprotein 37–8
    gene 37
tryptophan oxygenase 149
tubulin 10, 68, 87–8
    autoregulation of mRNA for 87–8
    degradation signals 89
    mRNA stability 84
tumour necrosis factor 149
tyrosine amino-transferase gene 103, 119,
    129

ubiquitin 125–6
undermethylation 119–20, 123, 127, 134

U RNAs 80
uteroglobin 149

variant surface glycoprotein (VSG) 37–8
viruses
    SV40 78, 131–2, 159, 161, 252
        large T antigen 156–8
    see also DNA viruses, retroviruses, RNA
        tumour viruses, named viruses
vitellogenin
    gene 129, 149
        activation of 102, 116
    mRNA stability 84, 90

Western blotting technique 4, 23

Xenopus laevis
    gamma actin 149
    hsp70 gene 159
    oocytes 174–6
    regulatory proteins 192
    5S genes 174–5
    Xfin protein 196–7

yeast
    amino-acid starvation 94
    genes
        a 33–4, 36–7
        α 33–4, 36–7
        homothallism (HO) 34–5
    heterothallic strains 33

homothallic strains 33
loci
    HML 33–4, 36
    HMR 33–4, 36
    MAT 33–4
mating type 32–7
    a 33
    α 33
    switching 33–6
mating type locus (MAT) 33–4
    homeodomain 189
    silencer elements 163–4
oncogene equivalents in 236
proteins
    a 189
    α 189
transcription factors
    ACE1 221–2
    ADr1 196–7
    GAL4 200, 209, 211–13, 215, 222–4, 226
    GCN4 94–5, 97, 204–5, 208, 211, 221,
        226–7, 242–4
    LAC9 200
    PPR1 200
    SW 15 197
yolk protein genes 58

Z-DNA 131–2
zinc finger motif 194–203, 226, 252
    multi-cysteine 198–203, 226
    two cysteine-two histidine 194–9, 226

1/10/91

Secondary studies

oLHβ

oPRL

→ ? Julie / Judy

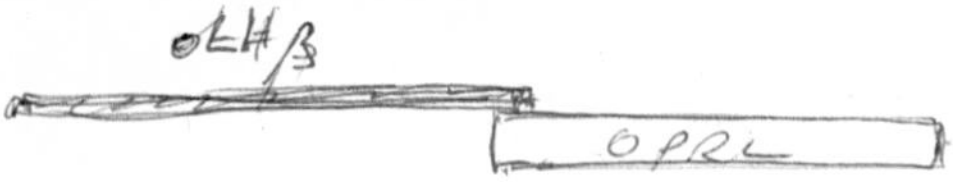

Why?
What is the question —

Sequestering
— control of in gonadotrophe
Secretion

— then oLHβ in lactotrophe

→ used bPRL (~~epithat~~)

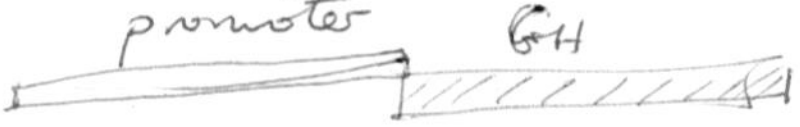

promoter   GH

Heap & ?
Sludge ?   →

↑ TRH ↓ GH   DA↓ GH .

reporter ?